Springer-Verlag Berlin Heidelberg GmbH

Springer Series in
MATERIALS PROCESSING

Series Editors: H. Warlimont   E. Weber   W. Michaeli

This series is focused on the science and application of materials processing as an essential part of progress in the materials field. It addresses researchers and process engineers alike. The scope of the series includes all classes of materials – metals, inorganic non-metallic materials, polymers and composites – in the form of bulk materials, thin films and layered structures as well as micro- and nanostructured forms. All aspects of materials processing from fundamental understanding to innovative strategies and methods of practical process implementation and control are covered. It is the aim of the series to provide comprehensive information on the science and application of leading-edge and well-established processing technologies.

Low-Pressure Synthetic Diamond. Manufacturing and Applications
Editors: B. Dischler and C. Wild

Purification Process and Characterization of Ultra High Purity Metals.
Application of Basic Science to Metallurgical Processing
Editors: Y. Waseda and M. Isshiki

Supercritical Fluid Science and Technology
Editors: Y. Arai, T. Sako, and Y. Takebayashi

Melt Blowing. Equipment, Technology and Polymer Fibrous Materials
By L. S. Pinchuk, V. A. Goldade, A.V. Makarevich, V. N. Kestelman

Epitaxy. Physical Foundation and Technical Implementation
By M. A. Herman and W. Richter

Series homepage – http://www.springer.de/phys/books/ssmp/

L. S. Pinchuk   V. A. Goldade
A. V. Makarevich   V. N. Kestelman

Melt Blowing

Equipment, Technology, and Polymer Fibrous Materials

With 105 Figures and 21 Tables

 Springer

Prof. L. S. Pinchuk
Prof. V. A. Goldade
Prof. A.V. Makarevich
V. A. Belyi Metal-Polymer Research Institute
of the National Academy of Sciences of Belarus
32a Kirov Street
246050 Gomel
Belarus

Prof. V. N. Kestelman
KVN International Inc.
632 Jamie Circle
King of Prussia, PA 19406
USA

Series Editors:

Professor Dr. H. Warlimont

Institut für Festkörper- und Werkstofforschung e.V.
Helmholtzstrasse 20, 01069 Dresden, Germany

Professor Dr. E. Weber

Materials Science and Mineral Engineering, University of California
587 Evans Hall, Berkeley, CA 94720-1760, USA

Professor Dr. Walter Michaeli

Institut für Kunststoffverarbeitung, RWTH Aachen
Pontstraße 49, 52062 Aachen, Germany

ISSN 1434-9795
ISBN 978-3-642-62785-9   ISBN 978-3-642-55984-6 (eBook)
DOI 10.1007/978-3-642-55984-6
Library of Congress Cataloging-in-Publication Data
Melt blowing: equipment, technology, and polymer fibrous materials / L. S. Pinchuk . . .(et al).
Berlin; New York: Springer, 2002. p. cm. Springer series in materials processing, ISBN: 3-540-43223-x
(alk. paper) 1. Plastics – Extrusion 2. Polymer Melting I. Pinchuck, L. S. (Leonid Semenovich) II. Series
TP1175.E9M45 2002   12701361

http://www.springer.de

© Springer-Verlag Berlin Heidelberg 2002
Originally published by Springer-Verlag Berlin Heidelberg New York in 2002
Softcover reprint of the hardcover 1st edition 2002

Typesetting by the authors. Data conversion and final processing by Peter Altenberg, Bremen
Cover concept: eStudio Calamar Steinen
Cover production: *design & production* GmbH, Heidelberg

Printed on acid-free paper   SPIN: 10791378   57/3141/mf - 5 4 3 2 1 0

Preface

One of the recently emerging techniques of fibrous materials production, melt blowing, consists of forming fibers from substances heated above their melting (crystalline) or glass transition (glass-like) point with further blowing by gas flow. The sprayed fibrous mass is then cooled to solidification either in a gas flow or upon deposition on the forming substrate.

Realized from polymers and then ceramics, the melt blowing technique has enriched materials science, engineering, and all commodity products by novel types of fibrous materials and products made from them with a unique combination of properties. The reasons for the popularity of melt blowing are the following.

The shape stability and strength of melt-blown materials and products are controllable technological parameters that depend on the diameter and the intensity of the adhesive interaction between fibers and the number of contacts between them.

The greater area of fiber surface in contrast to negligible clearances in between is the source of the uniqueness of melt-blown materials as systems whose properties are governed to a great degree by surface phenomena.

Dielectric materials manufactured by melt blowing are subjected to the rigorous effects of heat, deformation, and friction during processing which is accompanied by natural electrical polarization of fibers. The fibers are transferred into an electret state (an electret is a dielectric that preserves its electrical polarization for a long time), which makes melt-blown materials the source of a permanent electrical field.

The melt blowing technique creates new vistas for controlling the structure and properties of fibrous materials. At least four areas of control can be outlined.

First is the chemical composition of the material extruded into fibers. The second area is fiber transportation within the gas flow where the material is in a structurally sensitive state, either viscous-flow or viscoelastic. At this stage, it is convenient to modify fibers by chemical, physical, and biological methods. Third, the fiber diameter (from portions a micrometer to a millimeter) and a uniformity of the adhesion of dispersed components to the fibers can be adjusted to impart new functional properties to the material as a whole.

The fourth area is the texture of melt-blown materials and products that is determined by the mutual disposition and bonding of fibers to one another.

Development of a great variety of melt-blown materials has perceptibly impacted engineering domains and life as a whole. Following are some examples that confirm this fact.

Melt-blown materials can serve an ideal basis for biosorbents and biocatalysts in a number of biotechnological processes whose success influences their commercial prospects (biotechnology is a combination of industrial procedures using living organisms and biological processes in manufacture). Microorganisms immobilized on a fiber surface are easily accessible to reagents in liquid and gaseous phases. However, the shape stability of the fibrous carcass presents a mechanical barrier that separates microbial colonies from the environment. Weak and superweak physical fields generated by melt-blown materials also stimulate the vitality of microorganisms.

Melt-blown materials have opened new ways of solving problems in engineering ecology. Its methodology and tools require constant change in the range of filtering materials. Melt blowing technology has made it possible to simplify the problem of cleaning industrial wastewater and gas ejections, and to develop systems for entrapping petroleum products, organic solvents, heavy metal ions and to inactivate them biologically.

Recently elaborated melt-blown materials based on readily fusing glue compositions, also soft but preserving their shape lining, decorating, and other accessory materials have enriched light industry with novel techniques and products.

Melt-blown materials based on water-soluble polymers and their gels have formed the basis of a vast variety of medical, hygiene, cosmetic, and perfume products of a new generation without which modern civilization is unthinkable.

Unfortunately, despite almost a 50-year history, the melt blowing technique, for a number of reasons to be expounded further, is little known thus far. Until now, there has not been in any monograph in the literature that generalizes its objectives, means of attainment, and recent successes. What is more, the methodology, including its original tools, design of technological equipment, and instrumentation for implementing this unusual technology has not yet been elucidated. This book is the first publication where the physicochemical basis of the melt blowing technique is systematized, and fundamental flow charts, designs of the main joints, characteristics and fields of application for melt-blown materials are correlated. The authors have endeavored to describe precursors' works at length, even though the essentials of the book constitute investigations of their own completed at the Metal-Polymer Research Institute (MPRI) of the National Academy of Sciences of Belarus (Gomel, Belarus) with a Design Bureau and pilot plant.

The authors express their gratitude to MPRI's Director, Correspondent Member of NASB, Prof. Yu. M. Pleskachevsky for attention to this work;

Head of "Metal-Polymer" Co., Ph.D. A.I. Chernorubashkin, and Chief Designer of the company, A.V. Sikanevich, for permission to present data of the commercial use of melt-blown materials; Ex-Vice President of Korea Institute of Science and Technology (KIST), Prof. O.K. Kwon, Head of Tribology Center, Dr. U.S. Choi and Principal Researcher, Dr. B.G. Ahn of this Institute for cooperation in modernizing the melt blowing equipment, investigations, and the adoption of magnetic filtering materials in industry. We are also grateful to researchers, Ph.D. A.G. Kravtsov, I.Yu. Ukhartseva, and Yu.V. Gromyko for creative contributions to the experimental investigations of magnetic melt-blown materials. The authors are thankful to postgraduate S.V. Zotov and fellow-worker L.S. Pushkina for their invaluable service in preparing this treatise.

<table>
<tr><td>Gomel</td><td>Leonid S. Pinchuk</td></tr>
<tr><td>King of Prussia</td><td>Victor A. Goldade</td></tr>
<tr><td>January 2002</td><td>Anna V. Makarevich</td></tr>
<tr><td></td><td>Vladimir N. Kestelman</td></tr>
</table>

Contents

1. Introduction (Historical Review) 1

2. Melt Blowing Techniques 5
   2.1 Main Technological Procedures 5
   2.2 Modern Trends in Melt Blowing Techniques 10

3. Equipment 21
   3.1 Spray Heads 21
       3.1.1 Basic Designs 21
       3.1.2 Modified Heads 28
   3.2 Auxiliary Equipment 42

4. Structure of Melt-Blown Polymer
   Fibrous Materials (PFM) 53
   4.1 Major Structural Parameters 53
   4.2 Effect of Different Technological Regimes
       on PFM Structure 60

5. Specific Properties of Melt-Blown PFM 65
   5.1 Physicochemical Characteristics 65
   5.2 Electret Charge in Melt-Blown Materials 75

6. Fibrous Materials in Filtration Systems 83
   6.1 Efficiency of Filtration Systems 83
   6.2 Filtration Mechanisms 85
       6.2.1 Mechanisms of Particle Precipitation 85
       6.2.2 Surface and Depth Filtration 86
       6.2.3 Electrostatic Precipitation 89
       6.2.4 Precipitation and Coagulation in a Magnetic Field 91

7. Electret Filtering PFM 95
   7.1 Mechanism of PFM Polarization 95
   7.2 Capillary Phenomena 99

7.3   Production Process and Properties of Electret PFM 103
7.4   Applications 106

8.   **Magnetic Filtering PFM** 111
8.1   Background 111
8.2   Simulation of Magnetic Deposition in PFM 113
8.3   Theory versus Experiment 117
8.4   Magnetization of PFM 117
8.5   Magnetic Coagulation of Particles in PFM 121
8.6   Magnetic Capillary Phenomena 127
8.7   Serviceability of Magnetic PFM-Based Filters 132

9.   **Adsorptive and Microbicidal PFM** 135
9.1   PFM Modified by Porous Adsorbents 135
9.2   PFM as Adsorbents of Oil Products 137
9.3   Complex-Forming PFM 138
9.4   Adsorptive-Microbicidal PFM 143

10. **PFM as Carriers of Microorganisms** 147
10.1 Biofilters with Polymer Fibrous Biomass Carriers 147
10.2 Effect of Magnetic Fields on the Growth Processes
      of Microorganisms 155

11. **Other Applications of PFM** 161
11.1 Household Uses 161
11.2 Industry 165
11.3 Construction 168
11.4 Medicine 170
11.5 Packing 173
11.6 Protection of Products and Environment 175

12. **Ecological and Social Problems** 179
12.1 Solution of Ecological Problems 179
      12.1.1 Purification of Industrial Gases 180
      12.1.2 Wastewater Purification 181
      12.1.3 Melioration 182
      12.1.4 Oil and Chemical Sorbents 182
12.2 Regeneration, Utilization, and Burial 184
12.3 Economic Estimates 188

13. **Conclusion** 191

References 193

Subject Index 206

List of Abbreviations

AC – activated charcoal
AFM – atomic force microscopy
AOIA – automatic optical image analysis
BAPM – biological active polymer materials
BET – Brunour–Emmet–Teller theory
BF – barium ferrite
BLE – birch leaves extract
CC – coordination compounds
CCS – chemical current source
CNE – coniferous needles extract
COD – chemical oxygen demand
DEG – diethyleneglycol
DEL – double electric layer
DOP – dioctylphtalate
DSC – differential scanning calorimetry
EPR – electron paramagnetic resonance
ESCD – efficient surface charge density
FC – functional components
FE – filtering element
FM – filtering material
FPC – fibrous polymer carriers
FPF – fine-purification filter
HDPE – high density PE
HFC – high frequency current
IRS – infra-red spectroscopy
LDPE – low density PE
MF – magnetic field
MI – melt index
MPRI – Metal-Polymer Research Institute of National Academy
   of Sciences of Belarus
OM – optico-microscopic method
PA – polyamide
PAN – polyacrylonitrile
PE – polyethylene
PET – polyethylene terephthalate
PF – Petryanov's filter
PFM – polymer fibrous materials

PHC – polynitrogen heterocyclic compounds
PP – polypropylene
PTFCE – polytrifluorochloroethylene
PU – polyurethane
PVA – polyvinylacetate
PVC – polyvinylchloride
SEM – scanning electron microscopy
SF – strontium ferrite
TSC – thermally stimulated current
TSD – thermally stimulated depolarization
UV – ultraviolet
XPS – X-ray photoelectron spectroscopy
XSMA – X-ray spectral microanalysis

1. Introduction (Historical Review)

Melt blowing belongs, without doubt, to a newly developed industrial technique, though its onset dates back to the early 1950s. By that time, the world industry of chemical fibers already had at its disposal such highly productive and high-speed technologies as fiber formation from solutions and melts followed by linear extension [1, 2]. During the Cold War the U.S. administration showed an interest in producing of microfibrous polymer adsorbents intended for capturing radioactive particles in upper atmospheric layers which were signs of nuclear weapon tests intensively carried out in the U.S.S.R. during those years.

This order was fulfilled by National Aeronautics and Space Administration (NASA). One of its workers, van Wente, developed a piston extruder that forced a thermoplastic melt through a fine orifice. The melt stream, picked up by moving two flows of heated air at different speeds, was stretched out and dispersed. Upon cooling to a solid state, the fibrous fragments produced were placed on a substrate. By this method, 0.3 mm diameter fibers were produced. Van Wente encountered, however, a problem of instability in the technological process elaborated by him, which led to the formation of "pellets" between fibers. He could not solve this problem; so, at the end of the 1950s governmental financing of new technology ceased [3] because less labor-intensive technologies and less costly methods of detecting nuclear bursts had been developed by that time.

In 1965, Esso Research (at present Exxon Research and Engineering) revived this technology in search of new areas for polypropylene application. The problem of fiber formation without "pellets" was solved by a new design of an extrusion head. It substituted van Wente's one orifice head with a multihole head that had many orifices located in one line. The first products that Exxon developed by using this head were polypropylene battery separators and other electrochemical power sources, which have brought great success to the company. Probably, experiencing difficulties in increasing the output of melt-blown articles whose range and market were rapidly expanding, the company started selling licenses for the melt blowing technique. This was a decisive factor in further refining the method and perfecting the equipment.

The first among those who bought the license were V.R. Greis, James River, Kimberly Clark, and Minnesota Mining and Manufacturing. At present, this technology is being intensively developed at Hoechst AG, Johnson & Johnson, Procter & Gamble, Mitsubishi Gas Chemical, E.I. Du Pont de Nemours, Phillips Petroleum, Reifenhäuser GmbH, and other companies. It is natural that, companies, which received an exclusive right to use the licensed technology at a higher price did not rush to publish papers disclosing their knowhow. Until now, firms have been publishing mainly advertising information showing the advantages of the technique and manufactured products. This has become the main reason why there is, first, information "starvation" in all aspects of melt blowing realization and, second, practically a complete absence of publications about the effect of melt blowing technological regimes on the formation of melt-blown materials and products, as well as on their structure and properties.

From the end of the 1970s till the early 1990s, patents appeared, which proved the priorities of the firms mentioned in creating basic modifications of the melt blowing technique and major equipment designs [4]. The preceding years of long silence and the publications were signs that the secrecy period expired because the danger of losing the lead excited the firms more than probability of using data from patents published by competitors.

As for the former U.S.S.R., the history of the melt blowing technique started in 1970 when Academician V.A. Belyi (Director of MPRI in those years) invented a method for manufacturing coatings [5] by applying polymer melt simultaneously heated by IR radiation on a solid substrate. Curious enough that the argument, which forced an expert from a patent service to make a positive decision on this application, was the specific effect of pressure generated by a filament lamp light on molten polymer fibers when in dynamic contact with the substrate. They failed to obtain good quality coatings by this method, but this invention initiated an extensive complex of investigations at MPRI in developing polymer fibrous materials, technologies, and equipment for producting them.

This stage of melt blowing refinement is connected with the name of a talented Belarussian scientist, designer, and inventor V.P. Shustov. His creative ideas helped to further upgrade Belyi's achievements in manufacturing fibrous products displaying dimensional stability and to devise tens of technological modifications, new designs of aerodynamic heads enabling control of material texture, and new types of fibrous articles [6].

V.P. Shustov and his colleagues [7,8] extended the method to melioration which was intensively adopted in the U.S.S.R. during the years of "developed socialism". Ceramic drainage pipes fitted with melt-blown polyethylene connecting elements and polymer pipes with filtering melt-blown shells used for marshland reclamation were adopted almost all over the Soviet Union. A systematic approach to studying polymer melt dispersion gave birth to the scientific basis of the melt blowing technique.

Aerodynamic extrusion heads with a single hole proposed by Shustov were intended mainly for forming structural parts with a preset fibrous texture.

The production technique of melt-blown materials was further developed at the beginning of the 1970s in the All-Union Research Institute of Synthetic Fibers (Tver, Russia). A modification of fiber forming by a spinneret method complemented by interweaving fibers by airflow and thus forming a nonwoven fibrous cloth (the melt spinning process [9]) was named "aerodynamic forming" in Russia. The merit of this method is controllable production by a multihole die head and extruder as a whole. IR images and computer forecasting of mechanical properties of the formed melt-blown materials were used to study the formation regularities of sheet materials [10].

At the end of 1960s, the melt blowing method was used in Russia and Ukraine to produce fibrous materials based on ceramics: kaolin, basalt, refractories, and others [11]. This work continued in Belarus in the 1990s within the framework of the State Scientific and Technical Program "Triboengineering" and is underway today.

2. Melt Blowing Techniques

The melt blowing technique presents the unique possibility of varying the structure and properties of fibrous materials across a wide range, which attracts plastics manufacturers and designers. This variation might be realized at any stage of material molding, beginning with extrusion till cohesive bonding of cooled fibers on a substrate.

The basic technological procedures used to impart certain qualities to melt-blown materials will be considered in the first two chapters. Some original processes that create a diversity of physicochemical effects on materials, which conceal *know-how* prudently guarded by the developers, are also described here. As further narration will prove, they do not involve unpublished secrets, nevertheless, melt blowing is a rather freakish process and depends strongly on a number of factors, including the temperature and humidity of the granulate, the atmosphere and kind of the spray gas, and many others. The description of efforts to refine the melt blowing process will take into account all of these factors.

2.1 Main Technological Procedures

The **key diagram** of the melt blowing process is shown in Fig. 2.1. Polymer granulate *1* of composition C is processed in extruder *2* by regulating the rotational speed n of the worm and the temperature gradient T along the cylinder length. The melt is forced through the extrusion head dies *3* and is picked up by a gas flow from connecting pipe *4*. Molten fibers are elongated, transported in the gas-polymer flow *5* to the forming substrate *6* (revolving drum or mandrel, continuous band, immovable part, etc.), and deposited as a fibrous mass *7*.

The spray is controlled by regulating the melt temperature T_1 in the extrusion head and the spray gas pressure p. The fibrous structural formation of the melt-blown material is conditioned by technological factors; the most important among them are the nozzle-to-substrate distance L and the velocity v of the substrate motion relative to the gas-polymer flow. Their upper limits are bound by cohesive bond formation between fibers.

The criterion for optimizing technological regimes is the attainment of perfect service characteristics of melt-blown materials. They depend, all other

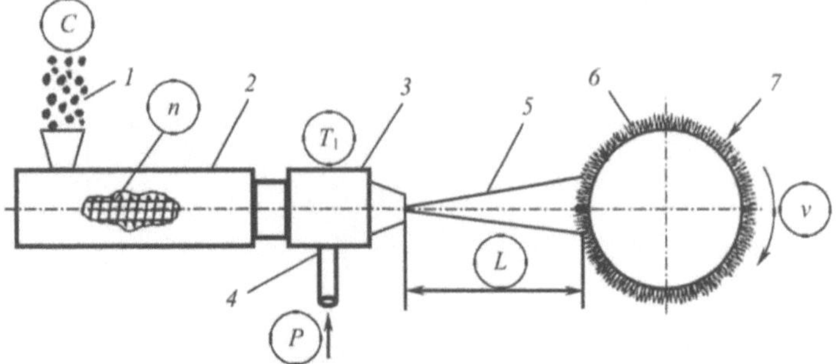

Fig. 2.1. Diagram of the melt blowing process: *1*, granulate; *2*, worm extruder; *3*, spray head; *4*, gas-feeding connecting pipe; *5*, gas-polymer flow; *6*, forming substrate; *7*, fibrous mass. The remaining notations are in the text

conditions being equal, on two structural parameters, fiber diameter d_f and density ρ (or porosity P) of the fibrous body.

The melt blowing technique has been embodied in multiple design modifications; among them the most wide spread are the following.

The **arrangement** of the forming substrate relative to the gas-polymer flow directed on it is selected from the criteria of technological losses upon formation of melt-blown materials with a given fibrous structure. As a rule, the flow is directed along the normal to the substrate. The basic schemes usually differ from those with horizontally [1] or vertically positioned substrates (Figs. 2.2a,b). Schemes with a vertically directed upward flow have not been used in melt blowing techniques. The flow can not be perpendicular to the substrate (c) which has been demonstrated elsewhere [2] in forming tubular elements by continuous withdrawal of the pipe and cutting it into sections (d).

The **number of extruders** serving gas-polymer flow generators is determined by the amount of components in a melt-blown material; yet, more than two or three extruders are rarely used simultaneously in practice. In Fig. 2.3, diagrams of the melt blowing process with two extruders are shown where the fiber flow is directed (a) along the normal or (b) at an angle to the substrate to govern either (a) layer by layer packing of fibers supplied from separate extruders or (b) mixing them on the substrate or in the gas flow. The flat forming substrate can be installed vertically.

Melt-blown articles can be formed from a mixture of fibers on rotating forming mandrels or by sequential deposition of layers. The latter variants are shown in Figs. 2.3c,d where the difference is in the coaxial or mutually perpendicular positions of the extruders.

Combination of the main melt blowing schemes with other procedures for woven and nonwoven material formation has given birth to a variety of original technologies [3]. Special technologies of fiber packing provide for control of nonwoven base texture in a melt-blown layer and the prevalence of

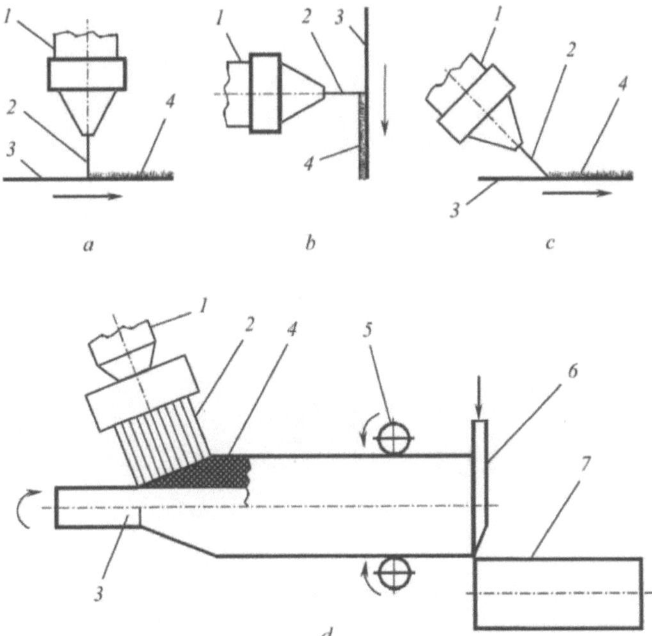

Fig. 2.2. A diagram of melt-blown material forming with fiber flow direction: (**a**) in the vertical plane; (**b**) in the horizontal; (**c**) and (**d**), in other planes with fiber deposition onto flat and cylindrical substrates. *1*, extruder with a spray head; *2*, gas-polymer flow; *3*, forming substrate; *4*, melt-blown material; *5* and *6*, drawing and cutting devices; *7*, tubular element

one kind of fiber on one side and of another kind on the opposite side of the cloth. Spunbonding techniques are closely related to melt blowing techniques. They employ high-voltage discharges in controlling relative fiber disposition, active gas media, mechanical fiber extension, and exposure to ionizing radiation. One more melt blowing analogue is radial extrusion of web materials through disk heads with spinneret holes along the cylindrical generatrix of the disk. It is often supplemented by the melt blowing technique to overlap web cells with thinner fibers.

There is also a large group of methods that use a combination of textile and melt blowing techniques. Although melt blowing ensures fiber cohesion at contact sites, the strength of the fibrous material is, in a number of cases, additionally improved by HFC welding, thermal and supersonic treatment of fiber contact zones, mechanical entanglement of fibers, including hydraulic methods, and by fastening melt-blown cloth using needle-punching.

Solid components are introduced into fibrous materials at different melt blowing stages (Fig. 2.4). Heat-resistant components, e.g., ferromagnetic powders, are processed together with the polymer binder in the extruder [4]. Often it is a necessary to fix solid particles on polymer fibers. Various

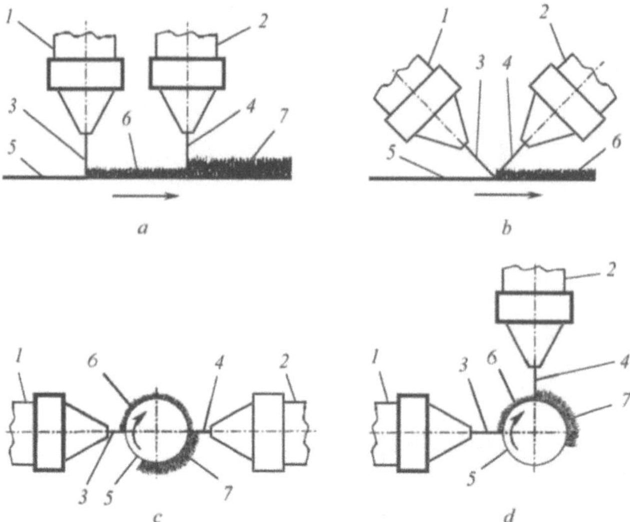

Fig. 2.3. Forming of melt-blown materials from two gas-polymer flows: (**a**) flows are normal to the flat substrate; (**b**) flows are perpendicular to the substrate; (**c**) and (**d**) fibrous layer application on a revolving mandrel in axial and mutually perpendicular flow directions. *1* and *2*, extruders; *3* and *4*, gas-polymer mixtures; *5*, forming substrate; *6* and *7*, melt-blown material layers

technological means are used to do this. The simplest is when particles are spilled onto the gas-polymer flow; this causes maximum powder losses. They can be reduced by counterflow mixing of polymer and modifying components (b) or over the forming substrate [5].

Powder particles can be satisfactorily fixed on fibers (c) by feeding powder on the deposited fibrous mass in a viscous-flow state or (e) by spraying polymer by solid particle aerosol. In diagram (d), the process uses three extruders and two powder flows. Due to this, the material achieved consists of three different polymers modified by two types of active . To reduce powder losses, the flow of modifier is directed between two gas-polymer flows. Mixing of components can be realized (f) on the mandrel surface at the moment of fiber deposition or (g) at an earlier stage before the flows reach the forming surface.

Modification of melt-blown materials by liquids is limited to a few variants. The most wide spread involves introducing components in a liquid phase into the extruded mixture (adjustment of composition C is shown in Fig. 2.1). In this way, plasticizers for the polymer binder are introduced and also, e.g., water that assists fiber dispersion in polymer melt spraying [6]. A fibrous mass can be settled on a liquid surface. In this case, fibers in a viscous-flow state actively interact with liquid components, in particular, with polymer solvents and plasticizers.

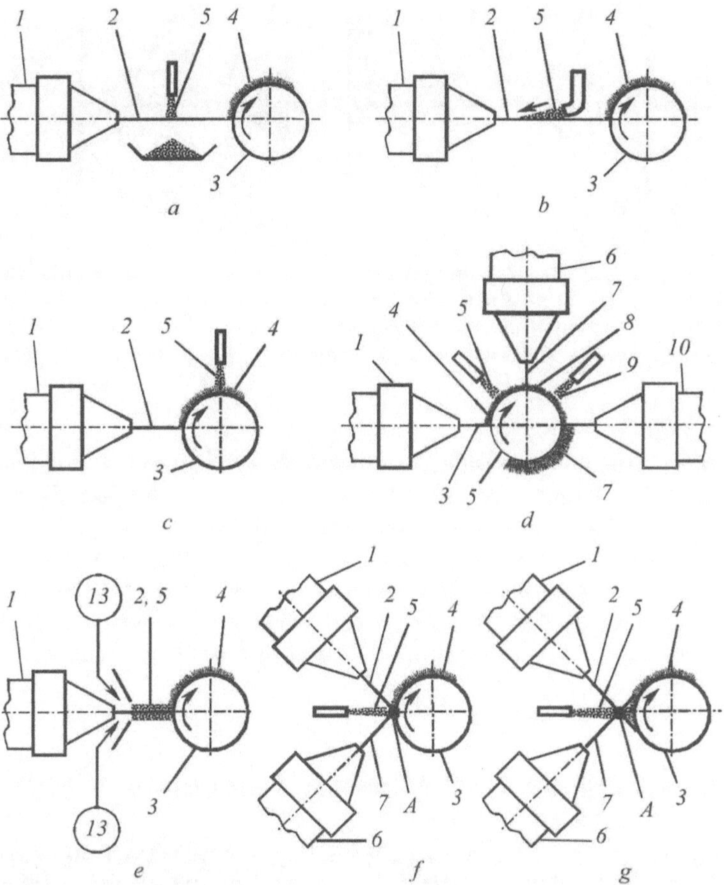

Fig. 2.4. Modification of melt-blown materials by solid-phase components: (**a**) powder feeding along the normal to the gas-polymer flow; (**b**) counterflow introduction of powder; (**c**) and (**d**) powder spilling on fibrous mass deposited by one and three extruders; (**e**) powder introduction into the sprayed gas flow; (**f**) and (**g**) powder flow directed between two gas-polymer flows mixed on a mandrel surface or before it. *1*, *6*, and *10*, extruders; *2*, *7*, and *11*, gas-polymer flows; *3*, rotating mandrel; *4*, *8*, and *12*, deposited fibrous mass; *5* and *9*, powder flows; *13*, gas-powder mixture generator; A, zone of mixing flows

Fiber treatment in gas media requires specific equipment, which isolates the process zone from the atmosphere (Fig. 2.5).

The gas-polymer flow is forced through a chamber limited by a hollow body where active or inert gases, sublimation vapor, or modifying liquid vapors are fed (a). Fiber treatment in gas media is often combined with modification by solid components. These procedures can be conducted in sequence (b) separated in time and space or simultaneously (c) when fibers are modified by solid particles in a gaseous chamber.

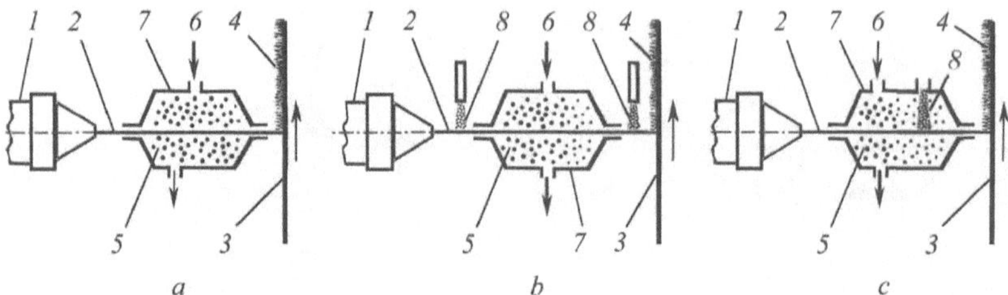

Fig. 2.5. Flow diagrams for processing melt-blown materials in gaseous media: (**a**) vapor-gas modification; (**b**) and (**c**) combined modification by gas media and solid-phase components, both sequential and simultaneous. *1*, extruder; *2*, gas-polymer flow; *3*, substrate; *4*, fibrous material; *5*, gaseous medium; *6*, gas source; *7*, body; *8*, powder flow

The flow charts previously considered admit additional effects on polymer fibers in the process of material melt blowing. These can be heat fluxes, electrical or magnetic fields, and electromagnetic radiation.

It is evident from the examples cited that melt blowing is conducted on special purpose unconventional equipment of unique design. Further, the structural and technological peculiarities of extrusion spray heads as well as some devices for fibrous mass deposition and finished melt-blown material removal is discussed.

2.2 Modern Trends in Melt Blowing Techniques

Achievements in melt blowing techniques attained during recent decades have defined trends in refining it in the near future. Most important among them are the following:

- productivity increase and reduced power consumption;
- elimination of polymer destruction and improvement of fibrous material strength;
- fiber modification upon leaving the extrusion head before solidification;
- the use of physical fields to control the process and impart additional functional properties to melt-blown materials;
- development of specialized technological procedures perfectly adjusted for forming certain types of products.

Increased productivity of the melt blowing method has always been a significant issue for the user because the efficiency of extrusion equipment is, as a rule, much higher than the carrying capacity of spray heads where the melt is forced through a limited number of small-diameter spinneret holes. This limitation brought about by different causes, including technological and designs, as well as consumer and ecological factors has conditioned the search for unusual ways of solving the problem. It consists of combining **polymer protection** against destruction with the **strength improvement** of fibrous materials.

The melt blowing process is usually run at temperatures corresponding to low-viscosity polymer melts. This is a double-edged weapon because of the danger of initiating thermal oxidative destruction of macromolecules when overheated polymer melt is sprayed in air. Biax–Fiberfilm Co. has patented a melt blowing process [7] that ensures two-step heating of the molten polymer. First, the extrudate is fed into an elongated channel of the spray head characterized by an insignificant temperature gradient across its length ($\sim 10°C/cm$). Then, it flows into the feeder subchannels. The melt remains in the elongated subchannels for less than 30 s after which is transferred into the heated spinneret channels from which is squeezed into fibers and sprayed by a hot air flow. In spinneret channels with a high temperature gradient, molten polymer remains for less than 2 s reaching a dynamic viscosity $\mu <$ 4.5 Pa·s at the moment it is sprayed. Such a heating regime produces conditions, first, for high productivity of the process owing to low melt viscosity and high fiber spraying velocity, close to the speed of sound. Second, this regime prevents exceeding the limits of intensive thermal oxidative destruction of macromolecules: the molecular mass of the fiber material is rather high, $M \geq 0.4\,M_0$, where M_0 is the molecular mass of the initial polymer.

The same Biax-Fiberfilm Co. has elaborated a process [8] realizing high productivity melt blowing in combination with heightened quality of the articles produced. The polymer melt is sprayed at $T < T_m + 50°C$ (T_m is the polymer melting point) with melt viscosity below 5 Pa · s and 150 to 300 m/s initial velocity. The regime requires a high degree of fiber orientation on a rotating mandrel. To control the shrinkage and density of fibrous tubular articles, the fibers deposited on the mandrel are blown off by a hot gas jet.

The spray gas temperature is a key factor determining energy consumption during the process and the quality of melt-blown products. The method already mentioned [1] presupposes melt spraying by gas flow at a temperature lower than that of the polymer in the head outlet. In this case, fibers are cooled down faster compared to the usual procedure. This is the basis for decreasing head to substrate distance and eliminating of pellets, fiber bulging, and other structural nonuniformities. It is obvious that the need to heat great volumes of the spray gas is avoided thus making the process more economical. Spraying with a cold gas protects the polymer from intensive thermal oxidative destruction even when air is used.

One of the ways of improving the strength of melt-blown materials is imparting an optimum fibrous structure to them. An original procedure of forming an anisotropic polymer body in porous materials [9] consists of stratifying the fiber on the forming substrate along a helical trajectory. To accomplish this, spraying is performed by a twisted gas jet. Polymer fibers are laid on the substrate in helical coils (Fig. 2.6) whose amplitude A and pitch H depend

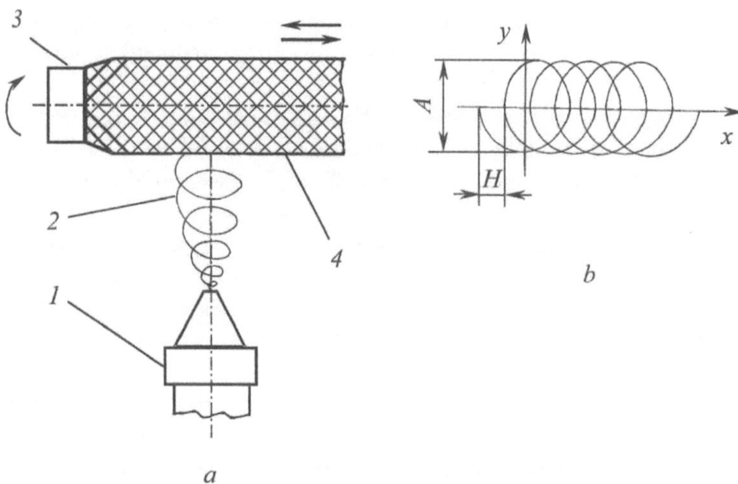

Fig. 2.6. Diagram of fibrous texture formation of a porous material: (**a**) arrangement of equipment; (**b**) trajectory of polymer fiber laying on the substrate. *1*, spray head; *2*, gas-polymer flow; *3*, forming substrate; *4*, polymer fibrous element

on substrate displacement parameters, as well as, the velocity and direction of the gas-polymer flow.

Fiber modification in a viscous-flow state is a convenient technological prerequisite for imparting unusual properties to melt-blown materials and products.

One of the first analogues of the technological method mentioned [10] involves simultaneous action of heat fluxes and the organic solvent vapors of decalin on the sprayed PE flow. This improves the strength of cohesive junctures between fibers and the density of the fibrous material.

A method of producing filtering materials [11] includes depositing a modifying layer adsorptionally active to sewage cations on the fiber surface. For this purpose, spray air is mixed with modifier vapors obtained by subliming of 8-hydroxyquinoline or *o*-aminobenzoic acid. To improve the of the modifying layer to the fibers, the mixture of air and modifier vapors is saturated with a polymer-solvent aerosol (e.g., vaseline oil) during material formation from polyolefins. The procedure is shown in Fig. 2.7.

Layers precipitated from the gaseous phase are thin and have technical potential. Fiber modification by a liquid-drop phase aerosol is more efficient. To produce filtering materials that can interact physically, chemically, or biologically with contaminants in the filtered media, the sprayed gas is mixed with an aerosol whose dispersed liquid bears active functions [12]. The following aerosols can be used with this aim:

- threebutyl phosphate which complexes with heavy metal ions that are typical sewage contaminants;

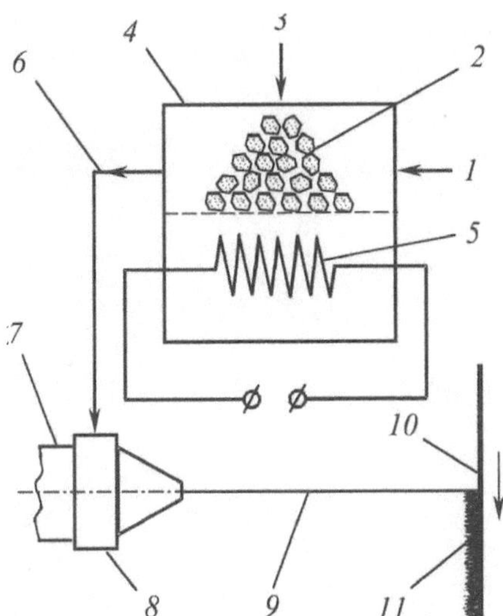

Fig. 2.7. Melt-blown material modification by subliming and depositing active components on the fiber surface. *1*, gas flow; *2*, modifier; *3*, aerosol flow; *4*, vessel; *5*, heater; *6*, aerosol flow with modifier vapors; *7*, extruder; *8*, spray head; *9*, gas-polymer flow; *10*, substrate; *11*, melt-blown material

- dithizone solution (complex) in tetrachloride carbon (a solvent for many thermoplastics);
- benzoic solution (microbicide) in vaseline oil (polyolefin solvent).

Solid modifier particles can be, in fact, brought into contact with polymer fibers in the viscous-flow state at any melt blowing stage (Fig. 2.4). The problem consists of realizing the process with minimum powder losses. One of the conditions here is prolongation of powder particle contact with the fibers. This condition is reflected in Fig. 2.4e, according to which the polymer melt is dispersed by a mixture of the spray gas and the powder modifier.

In Fig. 2.8, a scheme is given for one of the variants of the present technology [13] that provides for employing different powders as modifiers of polymer filtering materials:

- solid dielectric powders first subjected to electrical polarization before mixing with the spray gas;
- a mixture of powders with liquid glue aerosols, particularly, silicone-organic liquids in polyolefin processing;
- a mixture of powders, nitrogen, and bactericidal or aromatic substances.

The **use of physical fields** in melt blowing makes it possible to solve two groups of problems simultaneously: firs, addition of certain functional properties to fibrous materials and, second, control of the technological process itself.

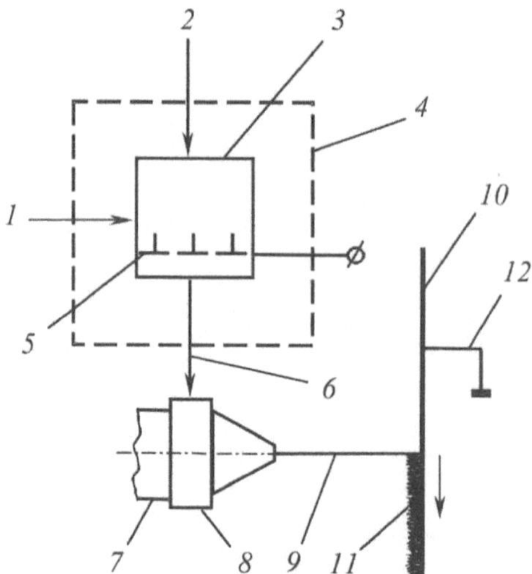

Fig. 2.8. A method for manufacturing fibrous nonwoven materials. *1*, *2*, and *6*, spray gas, modifier and flows of their mixtures; *3*, mixer; *4*, thermostat; *5*, electrode; *7*, extruder; *8*, spray head; *9*, polymer-modifier gaseous flow; *10*, substrate; *11*, fibrous mass; *12*, grounding grid

Electrical fields help to solve the problems in a very simple way. The method [14] described in Fig. 2.8 involves supplying an electrical voltage U from an outer source to electrode *5*. Solid particles fed into mixer *3* acquire under U a polarizing charge (predominantly unipolar) owing to which fiber polarization takes place during flow *9* shaping and transportation. Free charge carriers are removed to the grounding grid *12* as soon as fibers enter into contact with conducting substrate *10*. Electrical forces of the particle and fiber attraction to the substrate compact fibrous mass *11* and reduce its porosity. Simultaneously, particle losses during the process are reduced. This effect is intensified when an electrical potential is fed to the substrate whose sign is opposite to that of the polarized particles. When a potential unipolar to the fiber charge is supplied to the substrate, the porosity of mass *11* increases, and its density is reduced.

A new method of polarizing melt-blown fibers has been realized elsewhere [13]. In this method, fibers are obtained by forcing a melt through specially designed spinnerets in the extrusion head under an applied electrical potential. Because of this, the gas-polymer flow transported to the substrate carries its own electrodynamic field that interacts with the electrostatic field of the electrodes used to change the direction, density, velocity, and size of the flow cross section, thus adjusting it to the geometrical parameters of the substrate. The electrical charge on the fibers can be diminished by passing a gas-polymer flow through the aerosol cloud that serves as a fiber modifier. Grounding of the substrate brings about attraction of charged fibers by electrostatic forces and reduction of material losses during the process (see also Sect. 3.2, Fig. 3.22).

The method of manufacturing electret filtering materials [15] presumes grounding of the metallic forming substrate and application of a potential $U = (5 - 50)$ kV to a conjugated (with a 10 mm clearance) high-voltage electrode. The latter is fitted with a set of fine-pointed corona electrodes. Molten fibers are deposited on the substrate and transported together into the high-voltage electrode area to be cooled and transferred into the electret state during solidification in the U field (see detail in Chap. 7).

It is worthwhile using *magnetic fields* in the melt blowing process when forming magnetic-filled fibrous materials.

One of the firsts in this type of application [4] uses a sign-varying electromagnetic field to raise the degree of dispersion of a melt filled with ferromagnetic powder during fiber flow passage through the effective area of the field. A wide variety of technological means for controlling the structure of melt-blown material was found thanks to magnetic polymer filtering materials, many of which were created at MPRI NASB [16]. In Fig. 2.9 a scheme is presented to produce such materials. A polymer composition filled with a finely dispersed powder of magnetosolid barium or strontium ferrite is processed by a worm extruder and is squeezed through a spinneret as fibers.

At the outlet from the spray head *1*, circular electromagnet *2* creates the first texturing magnetic field whose intensity in the axial direction $H_1 \sim 200$ kA/m. Afterward, gas-polymer flow *3* passes through generator

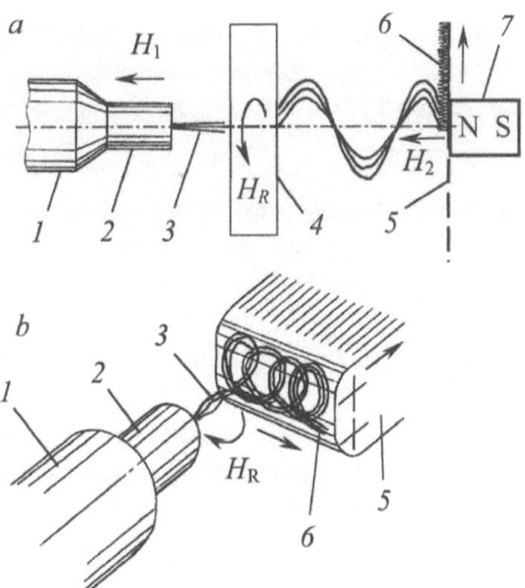

Fig. 2.9. A method of producing polymer melt-blown material: (**a**) general scheme; (**b**) scheme where fibrous mass stratifies on the substrate (see explanations in the text)

4 of electromagnetic field $H_R \sim 150$ kA/m rotating at $v = 3s^{-1}$ frequency and is directed onto mesh 5 where melt-blown mass *6* is precipitated in the second electromagnetic field $H_2 \sim 300$ kA/m. Its source is permanent magnet *7* located on the back side of the mesh.

Thus, the texture of magnetic melt-blown material is controlled in three areas:

- in the first texturing field H_1, particles of magnetic filler orient, in spite of short-term action, by the axes of slight magnetization along fibers;
- in the rotating magnetic field H_R, gas-polymer flow *3* deviates from its straightforward trajectory and juxtaposes on rotating and reciprocating mesh 5 as helical strips (Fig. 2.9);
- in the second texturing field H_2, fibers precipitate on the mesh predominantly along the normal which produces conditions to elevate the degree of magnetization of the melt-blown material by 10–15% in the outer magnetic field in contrast to the non textured one.

The magnetic field of the fibers filled with finely dispersed particles can be employed as technological and service factors in fixing coarse magnetic particles inside the fibrous matrix. Such a necessity arises in creating an intensity gradient in the filtering material in a magnetic field or when magnetosolid powders of coarse grist unsuitable for fiber filling are used as components. In this case, powders with particle size below 5 μm and, for spray gas, up to 200 μm are used for filling [17]. Consumption of coarse particles is controlled to form a concentration gradient of magnetosolid modifier in the filtering material. They are kept on the fibers by adhesive forces and magnetic attraction to filler particles encapsulated in the fibers. This can be attributed to magnetic saturation of particles introduced into the flow of spray gas and used as the filler owing to intensity $H \geq 8 \cdot 10^5$ A/m of the texturing magnetic field. This fact opens the possibility to

- form melt-blown materials with a stable fibrous structure because filler particles do not protrude outside the fiber contour and have rather low heat capacity which makes them unable to form "beads" and other nonuniformities in the fibrous body;
- raise the concentration of magnetosolid modifier perceptibly by fixing coarse particles between fibers not inside;
- lower consumption of deficient magnetosolid powders significantly by controlling their concentrations in the spray gas to generate a magnetic concentration gradient corresponding to the distribution of contaminants captured by the filter [17].

Melt blowing of magnetic polymer materials is hampered, to some extent, by the tendency of finely dispersed magnetosolid powders to form aggregates. This makes orientation of all particles by the axes of slight magnetization in the texturing field direction and, consequently, attainment of limiting values of material magnetic characteristics impossible. The method described in [18] involves extruding a mixture of polymer binder and magnetosolid filler under the simultaneous effects of static and variable magnetic fields. The static field has magnetic induction B = 16–18 mT, and the variable – magnetic induction gradient G = 700–800 mT/m and frequency ν =50 Hz.

Under the action of these fields, high local inhomogeneity of the magnetic field appears in the aggregate formed by a group of magnetosolid particles. As a result, each particle individually describes complex motions whose trajectory depends on mass, magnetic induction, and orientation of the particle and constitutes a combination of reciprocating and oscillatory-rotary displacements. Finally, internal friction in the polymer melt reduces, and the aggregate volume increases. Under an optimum correlation among parameters B, G, and ν and the time of field effect, an intensive destruction of aggregates occurs. Separated from the aggregate, particles are moistened by the melt averting further formation of aggregates.

The examples cited show that wide vistas are possible in using physical fields to control the production process and modification of melt-blown materials.

Specialized methods of manufacturing melt-blown articles are original technologies adjusted maximally for mass production of products of specific designs and designations. Articles such as bodies of revolution with intricate geometry are obtained by spraying fibers on a gas-permeable rotating mandrel inside which a vacuum or over pressure is created. Such a process scheme is shown in Fig. 2.10 [19].

Inside a perforated forming mandrel on which a fibrous mass is deposited, on over pressure $p \geq 0.1$ MPa is created. The flow of compressed air running through perforated holes reduces fibers packing density, elongates them on

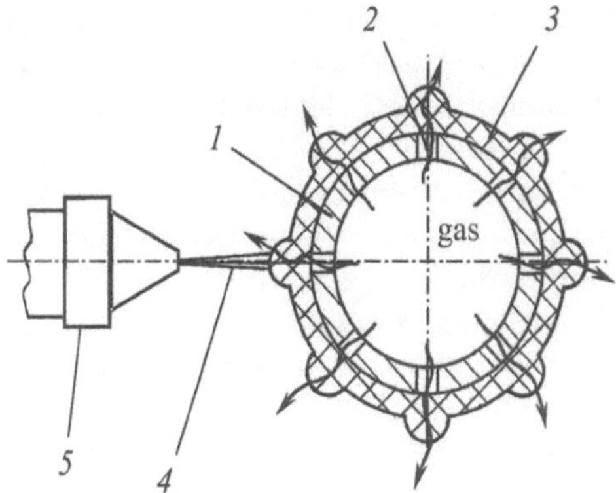

Fig. 2.10. Control of melt-blown article density. *1*, rotating mandrel; *2*, hole; *3*, fibrous mass; *4*, gas-polymer flow; *5*, extruder spray head

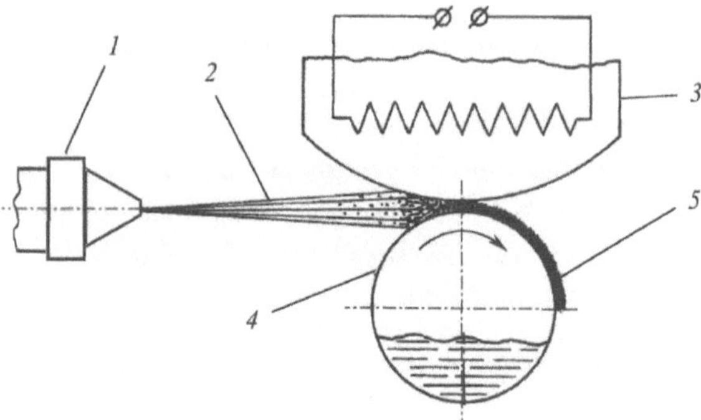

Fig. 2.11. A production scheme for sheet materials with controlled surface structure. *1*, extruder spray head; *2*, gas-polymer flow; *3*, heated plate; *4*, cooled drum; *5*, sheet material

local sites, and changes product geometry. A gas "pillow" between the mandrel and the product facilitates removal of the latter from the mandrel.

An original method [20] of producing melt-blown sheet materials that have a fluffy, porous surface on one side and a smooth surface impermeable to liquids on the other side provides for temperature gradient generation across the sheet thickness. A gas-polymer flow (Fig. 2.11) is directed into the clearance between the cooled ($T = 40°C$) drum and heated ($T \sim 200°C$) plate whose surface is coated with an antiadhesive.

When passing through the clearance, the fibrous mass is compacted and preserves its initial structure in contact with the drum, but while contacting the plate it partially fuses and becomes smooth and impermeable to media.

A special procedure has been proposed to manufacture drainage pipes with varying porosity across their thickness [21]. It consists of shifting the gas-polymer axis above that of the rotating mandrel during fibrous mass deposition. Due to this, part of the flow not overlapped by the mandrel forms a "tail" behind it from cohesively bonded fibers. The latter are additionally extended under the dynamic effect of the heated gas flow. During mandrel rotation, the tail is wound on it and fuses with fibers deposited on the mandrel that have a coarser structure. So, melt-blown pipes consisting of layers with different diameter fibers are formed during a single production cycle.

A special requirement imposed on a number of melt-blown articles is pre-set porosity of the fibrous matrix whose independent adjustment presents a serious technological problem. A method [22] helps to solve it by using two gas flows that fall on the forming substrate from both sides of the zone of deposition in the direction of substrate motion (Fig. 2.12). The gas temperature T is regulated, depending on the porosity gradient needed across the article thickness δ, i.e., with rising T and δ, the density of the article increases from

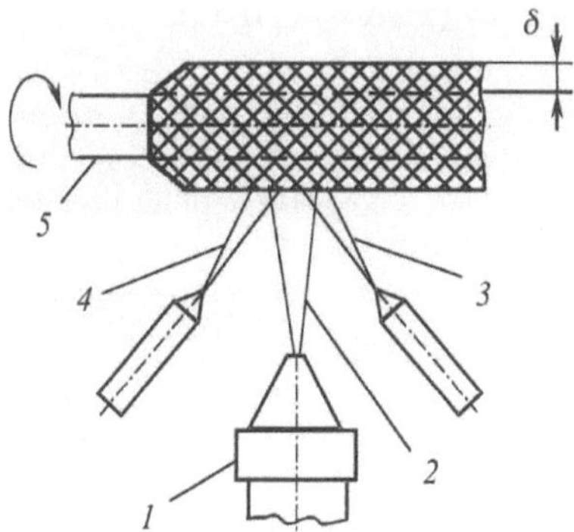

Fig. 2.12. A method layout with provision for controlled porosity of a fibrous matrix. _1_, extruder spray head; _2_, gas-polymer flow; _3_ and _4_, gas flows with controlled temperature; _5_, rotating mandrel

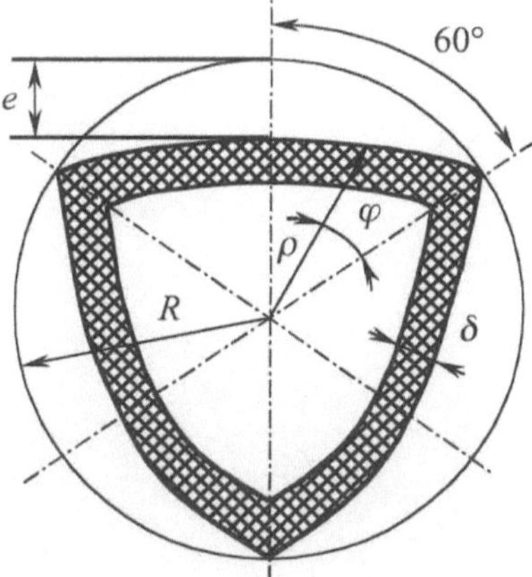

Fig. 2.13. Cross section of aerating element. See explanation in the text

the inside to the outer layer and upon reduction, vice versa, it diminishes in this direction.

Tubular aerating fibrous elements are often used to purify sewage with a high concentration of organic contaminants. They are insufficiently rigid upon bending, and during operation, air bubbles escaping from them coalesce and impair aeration efficiency. An aerating element that has a trihedral profile and curvilinear faces does not have these disadvantages (Fig. 2.13).

The radius vector of the profile is $\rho = R - e + e(1 - 3\varphi/\pi)^2$, where R is the rib radius, $e = \max(R - \rho)$, and φ is an angular parameter [23]. Such elements are manufactured by depositing polymer fibers on the mandrel of a corresponding profile. Under constant angular rotational velocity of the mandrel, the linear velocity of points on the surface is different, i.e., from $\omega \cdot R$ on its ribs to $\omega \cdot (R - e)$ on the middle of the face. As a result, the thickness δ and the porosity of the element walls are also different: δ is greater on the ribs compared to the faces, and the porosity is, vice versa, less. This distribution of thickness and porosity hampers the coalescence of air bubbles on the profile edges, and aeration efficiency improves.

The approaches in melt blowing techniques cited have been adopted in practice using original equipment and instrumentation, some of which will be covered in the following chapters.

3. Equipment

Equipment for melt blowing resembles facilities for coating more than machines for plastics processing, though it includes such traditional units as extruders fitted with high precision spray heads which are manufactured by special methods comparable to jeweler's art. That is why experiments connected with modification of precision elements of spray heads, receiving units, and other special equipment are highly expensive. So, the developers of melt blowing equipment highly rate the experience accumulated by them and this is, probably, the main reason why there is a scarcity of publications reflecting melt blowing problems in the scientific literature. Every design described in this chapter is based on *know-how* without which highly productive and reliable operation of equipment is improbable.

3.1 Spray Heads

Spray heads are devices located at the extruder outlet and that are designed to form fibers from the polymer melt. The fibers are further extended by gas flow, and the gas-polymer mixture is sent either on the forming substrate or into a medium where the fiber solidifies.

3.1.1 Basic Designs

Single-channel heads are the simplest type of spray devices. The first modifications of the device created by van Wente were equipped with a central channel through which polymer melt was forced, and a ring channel was located coaxially to the central one and connected with a gas source (Fig. 3.1a). The second type of head (b) involves gas flow through the central channel, and the polymer melt is forced through the ring channel embracing it that is connected with the working volume of the extruder [1].

An advantage of the first type is design simplicity and reliability; the second type of head has two times higher productivity compared to the other, all other conditions being equal [1].

Single-channel heads were refined in the 1970s and 1980s. Two main problems were solved during their modernization: control over the spatial

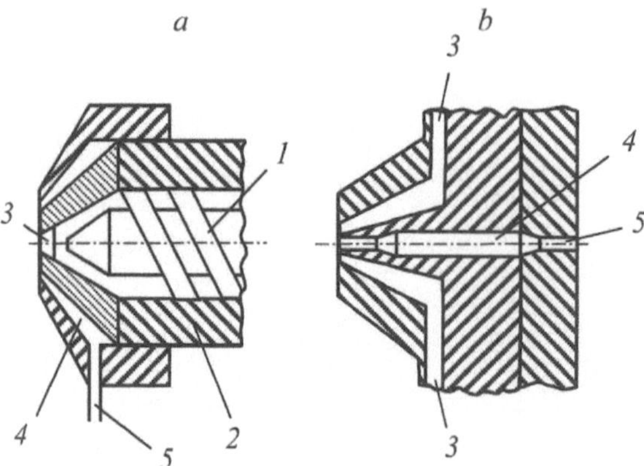

Fig. 3.1. Single-channel spray heads characterized by feeding either polymer melt ((**a**) type I) or compressed gas ((**b**) type II) through the central channel. *1* and *2*, worm and extruder cylinder; *3*, channel for polymer melt; *4*, cavity for compressed gas; *5*, compressed gas main

disposition of fibers during packing on the substrate and intensification of the polymer melt dispersion in the viscous-flow state.

The first problem was solved by imparting a helical trajectory to the gas-polymer flow. An example of the first type of head is device [2]. The channel for the melt is inside a cone detail fixed on the worm of extruder. Upon worm rotation, the cone that is fit with a system of grooves of a slide valve type gradually passes the gas flow through a set of holes around the channel. As a result, the gas-polymer flow acquires a spiral motion in the head outlet.

An analogous head of a more intricate design [3] is intended for applying melt-blown coatings on cylindrical articles, e.g., perforated pipes used in melioration. A swirler is installed inside a channel for spray gas delivery to initiate turbulence in the gas-polymer flow. Thus, the coating formed has many fiber welding "bridges" that make the coating stronger.

Dispersion of highly viscous melts can be intensified by employing vibration. Such a spray head is fitted with an electromagnetic vibrator [4] interacting with the extruder end piece. The latter serves as an electromagnetic armature and has a channel for melt release (Fig. 3.2). End piece vibration is transferred to the melt, excites pressure oscillatory pulses, and changes the sprayed gas direction. These intensify fiber stretching and breakage.

The main drawback of single-channel heads is their order of magnitude lower productivity than that when they are incorporated in extruders. Because of that, a key problem that faced to designers of melt blowing equipment was aligning of extruder and spray head efficiencies. The drawback was overcome by elaborating swirl (centrifugal) and multichannel heads whose basic designs are shown below.

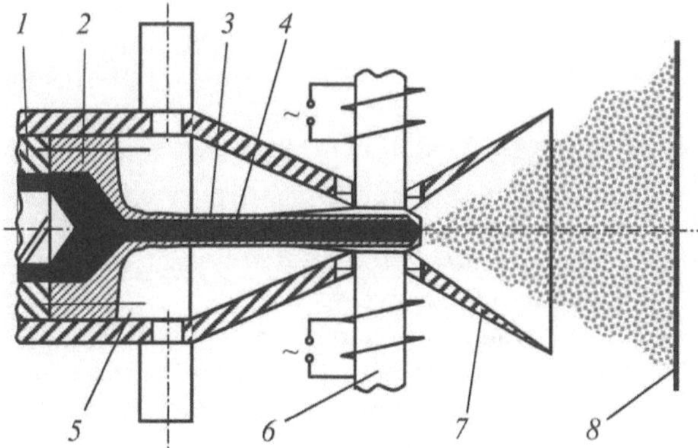

Fig. 3.2. Spray head with electromagnetic vibrator: *1*, extruder; *2*, head body; *3*, end piece; *4*, channel; *5*, gas-distributing chamber; *6*, electromagnetic vibrator; *7*, spray chamber; *8*, substrate

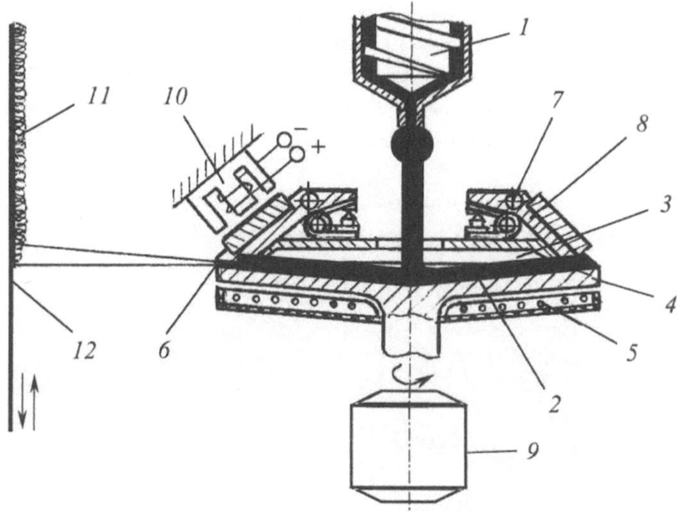

Fig. 3.3. Rotary spray head. See explanation in the text

Rotary spray heads disperse polymer melts by centrifugal force. They are close to single-channel heads because of the radial clearance for melt release and are proximate to multichannel heads because the clearance is usually divided into multiple radial outlet channels.

The operating principle of the rotary spray head can be considered by way of an example of a modification shown in Fig. 3.3 [5].

The outlet channel of extruder *1* is connected with clearance *2* between disks *3* and *4* where a temperature corresponding to the optimum melt viscosity is maintained by using heater *5*. Clearance *2* on the disk periphery breaks into open channels *6*. Spring-controlled sectors *8* overlapping from outside channels *6* are placed on disks *3* and revolve on axes *7*. Disk *4* is installed on the shaft of electric drive *9*.

When the electric drive is actuated, its system of disks starts to revolve, and the polymer melt moves into channels *6* in response to centrifugal force. When passing the effective zone of electromagnet *10*, sector *8* revolves under the action of electromagnetic forces of attraction around axis *7*, and overcoming spring resistance, opens outlet channels *6*.

The structure of the fibrous material is determined by the melt temperature in clearance *2*, disk *4* rotational velocity, and the sizes of channels *6*.

A similar design for polymer waste processing does not include any extruder [6]. It is made of two concentric chambers of a special shape installed on a vertical shaft, which are electrically heated. The polymer is loaded into the inner chamber, the heating system is switched on, and the shaft is set into rotation. The polymer melt is pressed against the chamber walls under centrifugal force and flows through special channels into the outer chamber from which it is then sprayed through the radial clearance. Substandard polymer containing impurities drops under the action of centrifugal and gravitational force down along the sloping wall in the inner chamber bottom and is also pulverized but through the lower radial clearance. In this way, spraying proceeds in two flows containing predominantly pure or substandard polymer materials.

An original device using inertial forces for polymer melt spraying [7] approaches rotary heads in its operating principle (Fig. 3.4).

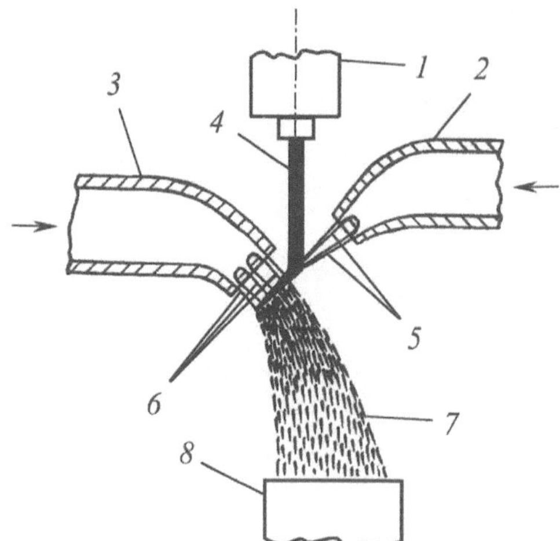

Fig. 3.4. A device for dispersing polymer melt film. See explanation in the text

The device consists of extruder *1* with a slot head and deflecting *2* and spraying *3* connecting pipes coupled to a compressed air main. Film *4* of the polymer melt issuing from the extruder head is deflected from the vertical direction by the dynamic action of a flat gas stream emanating from connecting pipe *2* that is parallel to the connecting pipe *3* face. A vigorous gas flow *6* emanating from it disperses the film, extends melt drops into fibers *7*, and transports them to tank *8* where they precipitate as a fibrous mass.

Multichannel spray heads are the main technological equipment for manufacturing melt-blown materials. Since their adoption, melt blowing techniques have become a flourishing industrial sector with a stable product range and market.

The basic designs of multichannel heads are not numerous. In Fig. 3.5, one can see a head with a wedge-shaped housing [8]. Housing *1* (the angle at the wedge vertex $\alpha = 30$–$90°$) is fitted with a set of holes *2* connected with extruder outlet *3*. The holes come out on the wedge face close to its edges to avoid dead zones. A 0.05-mm wide ledge between the hole edges and the face is believed to initiate dead zones. Cover plates *5* are installed on the head with clearances *4* relative to wedge planes. The clearances are connected to a compressed air source.

The head patented in 1972 by Exxon Research & Engineering Co. has become the prototype for several further generations of multichannel heads.

A similar design created in the 1980s (type II head) is shown in Fig. 3.6 [1]. The extruder forces the polymer melt into a heated chamber of the head where it is distributed among spinneret holes (a). The latter can be of various designs whose difference is in their location relative to nozzles connected to a compressed gas source (b–g).

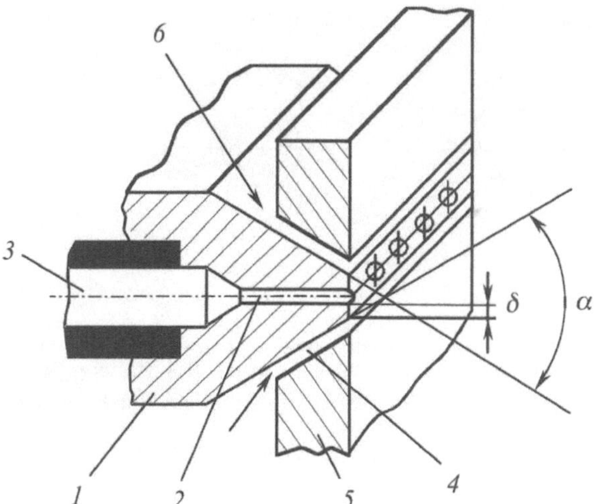

Fig. 3.5. Basic design of multichannel spray head. See explanation in the text

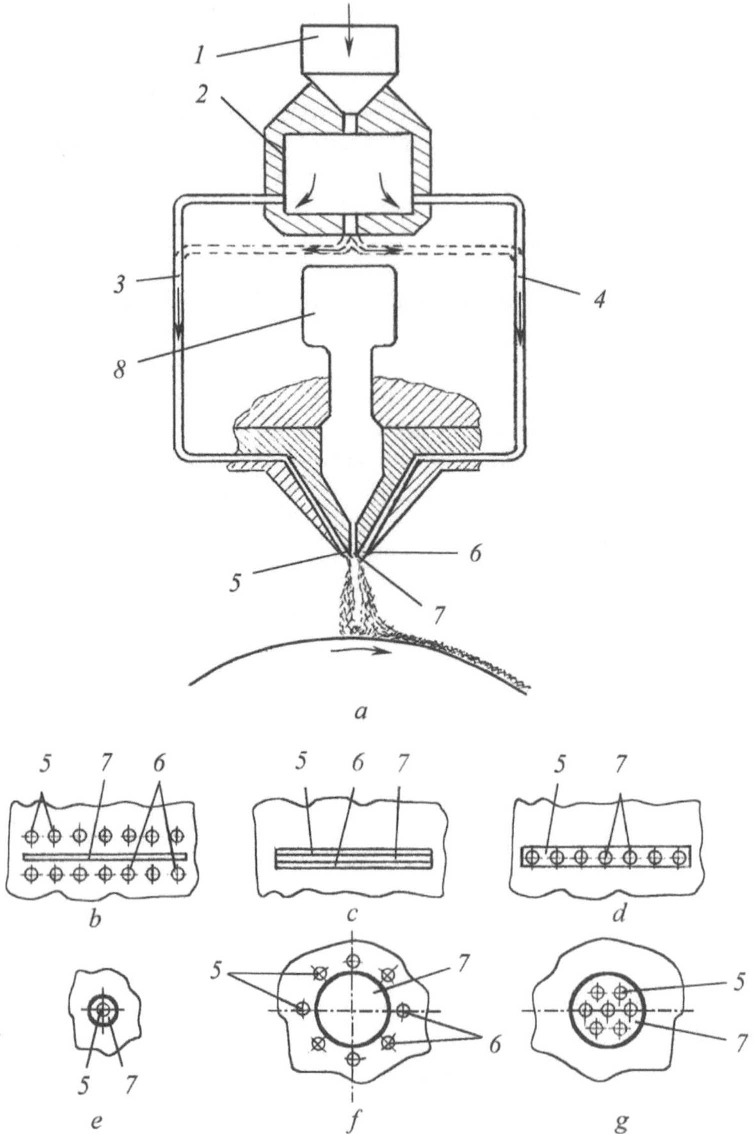

Fig. 3.6. A general diagram (**a**) of a multichannel head and modifications (**b**–**g**) of arrangements holes for melt release and compressed gas nozzles. *1*, extruder; *2*, chamber; *3* and *4*, channels; *5* and *6*, spinneret holes; *7*, gas nozzle; *8*, compressed gas source

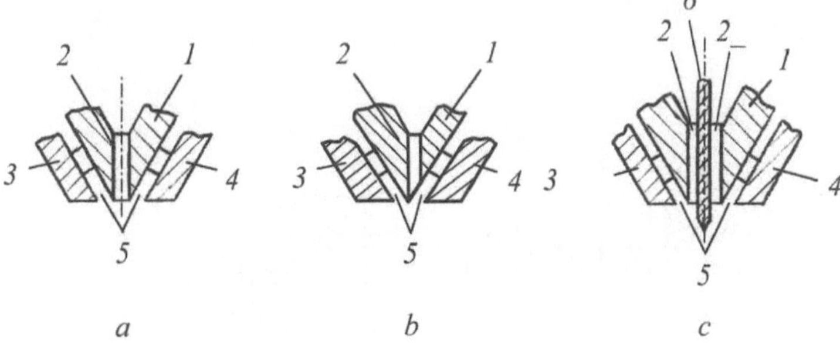

Fig. 3.7. Slot spray head. Slots: (**a**) symmetrical; (**b**) asymmetrical; (**c**) divided by a screen. *1*, housing; *2* and *2′*, slots; *3* and *4*, cover plates; *5*, clearances; *6*, screen

It is seen from the scheme (3.6 c, d) that channel *5* for melt release can be a rectangular slot which enables evening extruder and head productivity.

Such a design was first elaborated at Kimberly–Clark Corp. for type I heads [9]. The slot can be symmetrical (Fig. 3.7a) or asymmetrical (b) relative to clearances *5* through which gas streams are flowing. It may be divided into two channels by flat screen *6* (c). Due to this, simultaneous processing of two polymer materials has become possible; each is squeezed through flat channels on either side of the screen.

However, it was not the slot head that assured the success of melt blowing technology. Heads with numerous spinneret holes have played a decisive role in polymer melt release.

How to balance melt pressure which drives apart large size spinnerets has become a serious problem in designing heads with a multihole spinneret. A typical design, which outlined directions for solving this problem, is shown in Fig. 3.8 [10].

Head (a) includes a housing made of two rigidly connected parts with a spinneret fixed to it that consists of two halves *2* and *3* and a pair of blocks *4* and *5*. There are numerous holes in spinneret *6* (*4* to *16* holes per cm with a 0.25–0.60 mm diameter) connected through channels *7*, *8*, and *9* with the extruder working volume. Perforated plate *10* at the spinneret inlet is intended to apportion the melt evenly among holes *6*.

The spinneret is secured on the housing by bolts *11* and *12*. Tightening them seals the junctions of the polymer melt channels and brings about stresses preventing the spinneret lips from opening (described below). Blocks *4* and *5* form cavities *13* and *14* together with the housing where compressed gas is fed through connecting pipes *15* and *16*. Gas flowing through channels *17* and *18* transports fibers squeezed through the spinneret to the receiving unit.

The main drawback of multichannel spinneret operation is that the spinneret is pushed open under extrusion pressure. As a result, the melt in the central part of the spinneret is not squeezed through holes *6* but through a

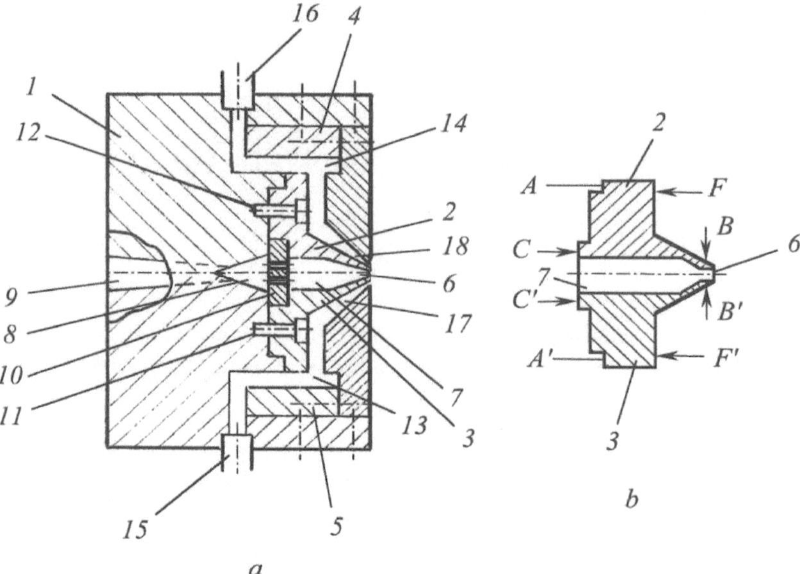

Fig. 3.8. Principal diagram (a) of the head and (b) distribution of forces in a spinneret at head installation. See explanation in the text

slot formed when the spinneret lips are forced apart. Naturally, the quality of melt-blown materials drops. In this connection, grooves were made on the spinneret bearing surface, so that when tightening bolts *11* and *12*, planes A and A' are first brought into contact with housing *1* (Fig. 3.8b). Tightening forces F and F' create bending moments relative to A and A' which generate, in their turn, oppositely directed and equal forces B and B' counterbalancing the extrusion pressure. The spinneret dimensions and those of the coupled part of the housing should produce conditions so that forces B and B' appear earlier than forces C and C', thus ensuring melt channel sealing. The range of compression stresses generated in response to forces B and B' (from 7–70 to 138 MPa) provides for normal operation of heads with a multichannel spinneret.

3.1.2 Modified Heads

Control over temperature in spray heads is an actual problem whose solution determines process productivity and the quality of melt-blown products.

Kimberly–Clark Co. has suggested original methods for stabilizing temperature in heads. A spray head [11] contains heat-insulating elements, which prevent spray gas heating on contact with the heated spinneret. With this aim, a metallic spinneret is furnished with $\delta \sim 0.5$ mm thick coatings of silicon ceramics, e.g., porous boron silicate, applied to surfaces washed by a

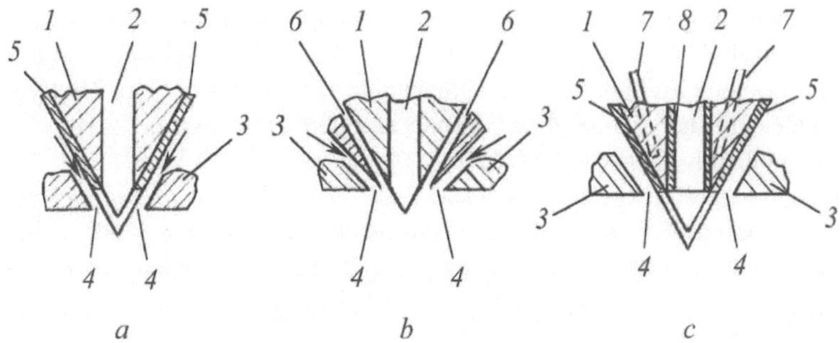

Fig. 3.9. Regulation of spray head temperature using: (**a**) heat-insulating coatings; (**b**) air gaps; (**c**) heating elements. *1*, spinneret; *2*, outlet for polymer melt release; *3*, gas flow guide; *4*, gas nozzle; *5*, heat-insulating coating; *6*, air gap; *7*, electric heater; *8*, electrically heated coating

gas flow (Fig. 3.9a). Heat insulation can result from an air gap (b) between the spinneret and the wall of the gas channel. In some cases, the spinneret temperature is controlled by band heaters, including electrically heated coatings (c).

All of these measures are intended to reduce the distance L from the spinneret to the forming substrate (see Fig. 2.1). In the majority of cases, melt-blown material is obtained with satisfactory properties when fibers have a diameter $d_f \sim 10\mu$m and length $l \sim 5$ mm. For this purpose, gas flow is sprayed at $T = 38\text{--}40°$C, and $L = 200\text{--}300$ mm. The proposed head reduces L to 150–180 mm without any special cooling of the spray gas.

When a wide (more than 1 m) melt-blown fabric is manufactured, a problem arises of uniform polymer melt supply to spinneret holes situated at different distances from the central channel through which melt is fed to the head. The problem is solved by a multihole-spinneret head [12] that has a full-length hollow chamber where the heated gas ($T \sim 290°$C) is delivered by a compressor. It heats the pipe connecting the working volume with the feeder through which melt ($T < 290°$C) is supplied to the spinneret holes, the spinneret, and the head body. Gas flowing from the nozzles (at a pitch $h = 1\text{--}4$ mm along the head length) blows out polymer fibers squeezed from the spinneret.

A number of computer programs have been elaborated to calculate gas and heat flows in such heads, e.g., Fluent (suggested by Fluent Inc.). The initial parameters are the mass of the head, the gas flow, and the polymer melt characteristics and melt blowing efficiency. At an optimum distribution of the cross-sectional area along the gas channels, the gas flow velocity ($v = 3\text{--}6$ m/s in the inlet) will grow with the flow and reach 24–57 m/s in the chambers transporting gas further to the nozzles. As a result, the pressure drop of the gas between the head inlet and the nozzle is only $\Delta p = 3\text{--}10$ kPa, and the temperature difference ΔT between the central part and head edges does not exceed a few degrees. This is much better than traditional multichannel heads with electric heating (Δp =35–50 kPa, $\Delta T \sim 40°$C). Heat exchange between the housing and gas is approximately five times higher than between the head walls and the atmosphere. When the head is insulated, its surface temperature will not exceed 40°C [12].

With the dissemination of the method, *the amelioration of the technical and economic parameters* of melt-blown materials and their production process is becoming a still more urgent problem. Implementation of the process at high melt viscosity consumes large amounts of spray gas ($q > 20$ g of gas per g of product) under a fiber spray velocity less than sonic. Viscosity reduction leads to a perceptible lowering of air consumption and attainment of optimum process productivity but is connected with macromolecular degradation. The latter results in impaired deformation and strength characteristics of fibers and the product as a whole.

Heads have been created by the Biax–Fiberfilm Corp. to overcome this contradiction. The head [13] design is based on the following requisites. Polymer diskharge through a spinneret hole should be rather low to form superfine melt-blown fibers. The fiber diameter is $d_f = \sqrt{4Q/\pi\nu}$, where Q is the polymer mass flow rate through the hole, and ν is the fiber spray velocity, which is below sonic velocity. This is the reason why Q values at $d_f \sim 1$ mm are very low. To achieve high process productivity in such conditions, a head is required with a great number of spinneret holes.

The head structure is shown in Fig. 3.10. Pipe *1* connects extruder working volume *2* with cavity *3* in plate *4*. A few rows of tubes *5* whose through holes connect with cavity *3* are secured in the plate (silver soldered). The tubes pass through collector *6* connected with compressed air main *7* and come out on the head face through holes *8* in plate *9*.

The bottom view of plate *9* is shown in Fig. 3.10b,c, where air streams are emanating from the collector parallel to tubes through cavities in the corners of square or triangular holes.

The collector is equipped with manometer *10* and thermocouple *11*. Flowmeter *12* and heater *13* are installed on tube *7*. The polymer mass temperature in pipe *1* is controlled by thermocouples *14–16*. Polymer pressure pickup *17* is located in cavity *3* near the inlet to the tubes. Bypass pipe *18* and valve *19* are intended to shunt the polymer flow through the head and to reduce the flow rate in tubes *5*. Valve *19* is used to regulate heat exchange and melt temperature in pipe *1* and the tubes.

A variant of arranging holes in plate *9* for air outlet from the collector is shown in Fig. 3.10d. The holes are shaped by mesh *20* and airflows *21* emanating from clearances between the wire, and tubes disperse the polymer fibers, which are squeezed through the tubes.

A physical model has been proposed to describe the heat exchange processes in this head [13]. The Fourier equation for heat exchange in cylindrical tubes is as follows:

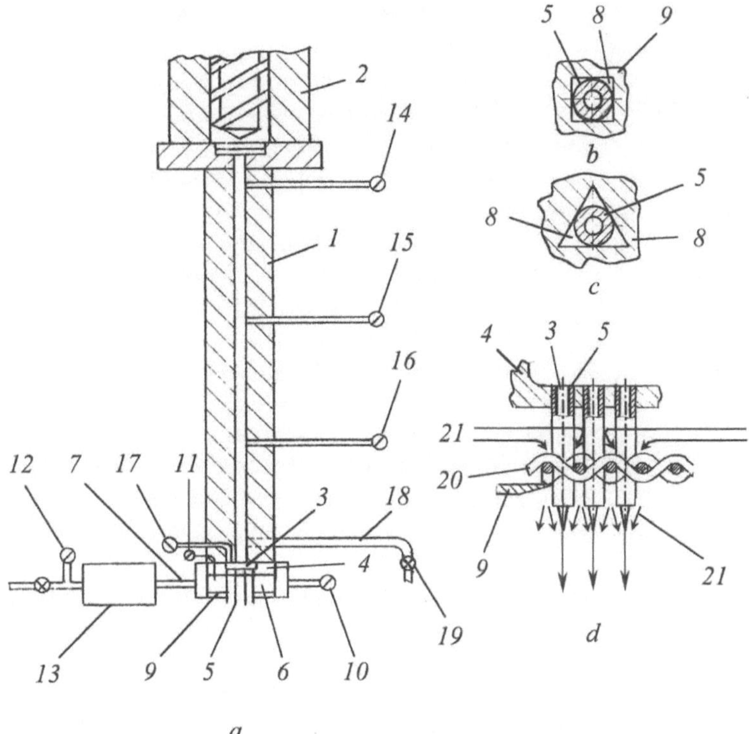

Fig. 3.10. A spray head for forming micron size fibers from polymer materials with a reduced degree of destruction: (**a**) general view; (**b**) and (**c**) head face; (**d**) a view of holes for air outlet through a metal mesh. See explanation in the text

$$\frac{d^2T}{dr^2} = \frac{1}{a}\frac{dT}{dt} ,\qquad\qquad (3.1)$$

where T is temperature,°C; r is tube radius, cm; t is time, s; a is $\lambda/(c \cdot \rho)$ – thermal diffusivity, cm^2/s; λ is thermal conductivity, cal/(°C·s·cm); c is heat capacity, cal/(g·°C); and ρ is density, g/cm^3.

Although (3.1) describes an ideal system and does not account for local temperature changes, boundary conditions or pipe cross section, it approximates kinetic processes and design peculiarities perfectly. The dimensionless relation at/r^2 (Fourier number) characterizing immovable (fixed) systems can be used to describe the polymer flow in the spinneret channels. If $v_p = l/t$ and $A = Q/v_p$, then $t = A \cdot l/Q$, where v_p and t are the polymer flow velocity and the time for passing through a channel of length l, A is the channel cross section, and Q is the polymer flow rate (volume/time) through A. Consequently, $at/r^2 = \pi a l/Q$. For noncylindrical channels, $r = 2A/P$, where P is the perimeter of the channel wetted by the polymer.

As can be seen, the polymer melt fed from the extruder into the head [13] passes, sequentially, the first heating zone (pipe *1*) characterized by a low temperature increment, and the second zone (tubes *5*) where it is quickly heated to high (370–400°C) temperatures. Low melt viscosity ensures minimum residence time of the polymer in the tubes, preserves it from intensive destruction, and ensures conditions of high (equal to sonic) spray velocity which guarantees high head productivity.

A multichannel spray head [14] outdoes the previous head in the efficiency and quality of melt-blown products and prevents the formation of pellets in the fibrous structure. It was noticed during operation [13] that the main factor in producing high-quality fibers was the central location of tubes relative to the spray gas flows in the head. It was proved that round holes for the gas were more advantageous than square or triangular holes (Fig. 3.10b,c). If the tubes are arranged in more than four rows, then to attain even formation of fibers, it is worthwhile to lengthen them up to $l \sim 6d$, where d is the tube inner diameter. The tubes should be at sufficient a distance from each other to avoid sprayed fiber interaction with the neighboring gas flow blowing fibers from another tube.

A head design based on these principles is shown in Fig. 3.11. Cavity *2* in housing *1* supplies polymer melt from the extruder. It connects with holes in tubes *3* installed in plate *4* at a distance $s \geq 1.3d$ from one another (d is the inner diameter of the tube). Beneath is chamber *5* whose height H is *know-how* in the patent [14]. Compressed gas from connecting pipes *6*, cavities *7*, and slots *8* (1–1.5 mm wide) is fed into the upper part of the chamber. Inside chamber *5* are tubes *3* 290–300 mm long joining a set of gas plates *9–12*. The latter are tightly secured coaxially in plate *9* with clearance to holes *13* and *14* in plates *11* and *12*. Separator plate *10* forms cavity *15* of $h \geq d/2$ between plates *9* and *11*, and the total thickness of the set of plates *9–11* is not less than $10d$.

The gas issues from chambers *5* through holes *16* in plate *9* into cavity *15*, then evenly flows around the nozzles, leaves the head through holes *13* and *14* (in holes *14* gas expands), and disperses fibers *17* which are forced through the tubes.

Variants of hole arrangements in the gas plates to raise the productivity of the head and obtain micron size fibers and uniform fibrous structure are shown in Fig. 3.11b–e.

Strengthening of melt-blown articles is an actual objective retained in various circumstances (production facilities, designation of products, processed materials, etc.). It is essentially solved by designing heads that can control mechanical and cohesive bonding of fibers during their travel from the head to the substrate and during packing.

An example of an extrusion head [15] distinguished from the basic design [8] by the location of spinneret hole axes at an angle $\alpha = 2$–$10°$ to the spray direction is shown in Fig. 3.12a.

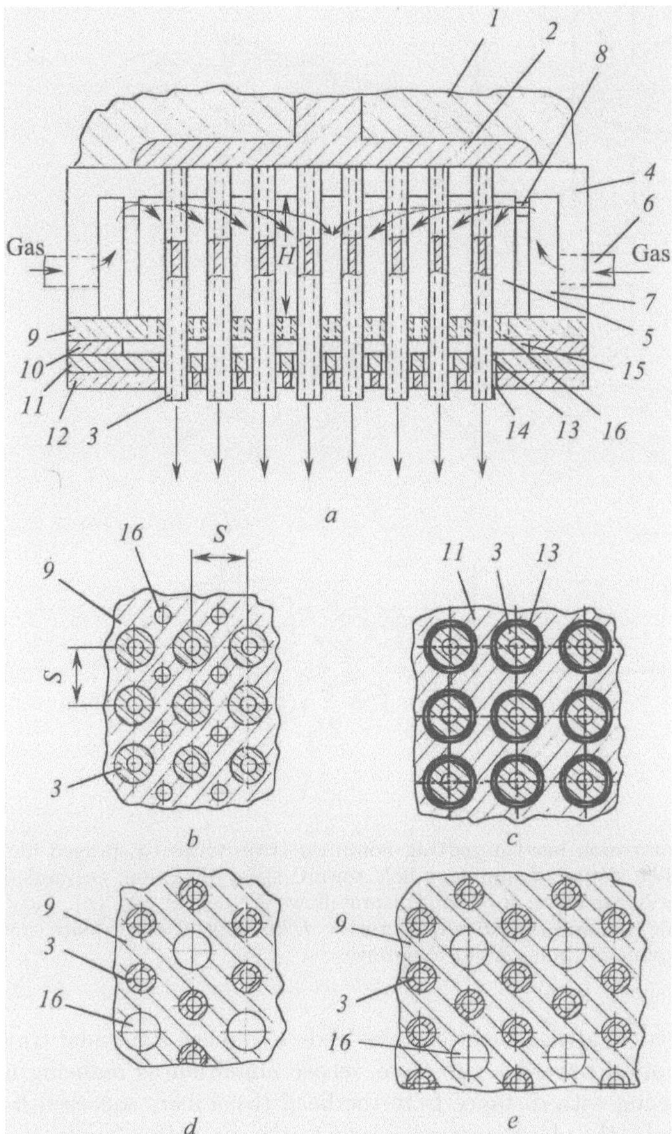

Fig. 3.11. A spray head with multiple rows of spinneret holes to form homogeneous melt-blown materials: (**a**) general view; (**b**) and (**c**), bottom view of plates *9* and *11*; (**d**) and (**e**), modifications of plate *9*

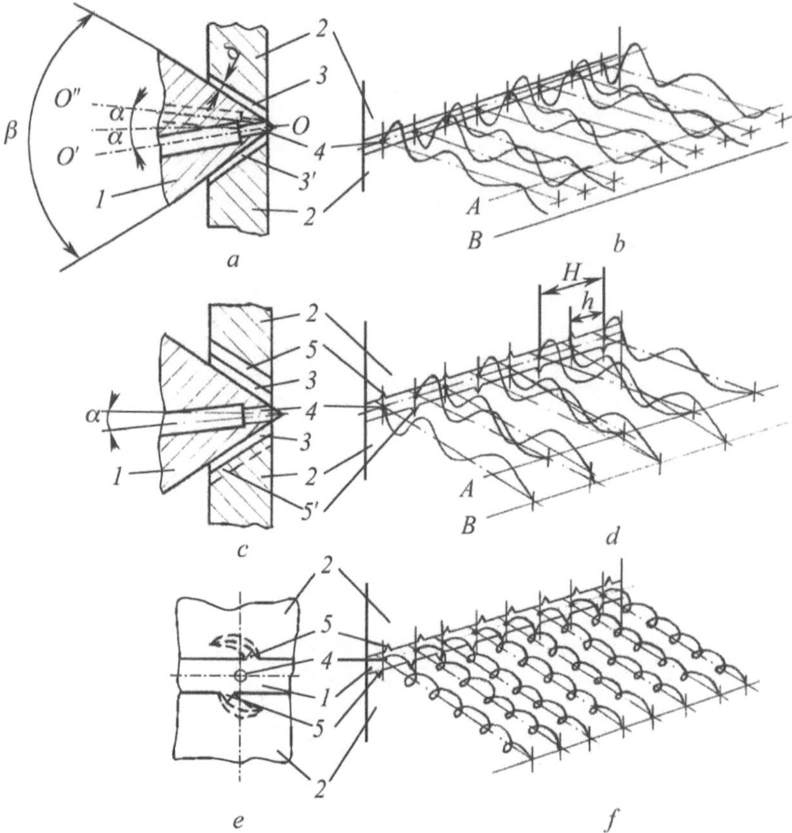

Fig. 3.12. An extrusion head imparting nonlinear trajectories to sprayed fibers: (a) and (b), unlike slopes of spinneret hole toward spray direction; (c) and (d), asymmetry of spray gas flows; (e) and (f), spray flows are helical. (a), (c), and (e), head designs; (b), (d), and (f), fiber trajectories. *1*, housing; *2*, cover plate; *3* and *3'*, gas gaps; *4*, spinneret hole; *5* and *5'*, grooves

An asymmetrical arrangement of holes leads to a quasi-sinusoidal trajectory of fiber motion toward a substrate, whose amplitude is reducing and period is increasing with distance from the head (b). Fibers squeezed from odd holes will cross the plane perpendicularly to the spray direction in points along line A, and fibers from even holes – along line B. The greater angle α, the longer is the distance between lines A and B, the longer the fiber loops laid on the substrate.

In Fig. 3.12c, a variant of the head is illustrated with asymmetry of spray gas flows relative to a spinneret hole. Grooves *5* and *5'* connect with clearances *3* and *3'* that join with the housing cover plates. The groove is made in only one cover plate near each hole at a pitch $H = 2h$, where h is the center-to-center distance between holes. During operation (c), gas flow power from gap *3* and groove *5* is greater than that of the gas flow from the opposite gap *3'*. This is the reason why trajectory (d) of fiber squeezed from hole *4* deviates during motion with the gas flow to the opposite groove *5* side. Fiber trajectories cross the plane normal to the spray direction in points on lines *A* and *B*. The larger the cross-section of grooves *5* and *5'*, the greater the distance between *A* and *B*.

Such a head acquires unusual features if grooves in both cover plates are helical to hole axes (e). In this case, leaving grooves *5* and *5'*, the gas (f) creates a concentric flow which draws fibers into a spiral motion thus increasing the probability of contacts and splices with neighboring fibers.

The heads imparting nonlinear trajectories to sprayed fibers are the means of simple and efficient textural and strength control of melt-blown materials.

Analogous problems can be solved by heads [16] that issue pulse gas flows. Their design is conventional (Fig. 3.13a) except for channels *7* and *8* in cover plates adjacent to spinneret holes and connecting with gas distributor *9*. In operation, gas flows from the holes (indicated by letters) in the cover plates and diverts sprayed fiber trajectories from their direction along the OO' axes of the spinneret holes.

When gas flows from upper holes (b, d, f, h, j, l, n, p...) and lower holes (a, c, e, g, i, k, m, o...) are switched on in turn, fiber trajectories acquire a congruent sinusoidal shape (b). They occupy vertical planes, which cross spinneret hole axes. The sinusoid amplitude depends on pressure, and its period on the time of gas flow switching.

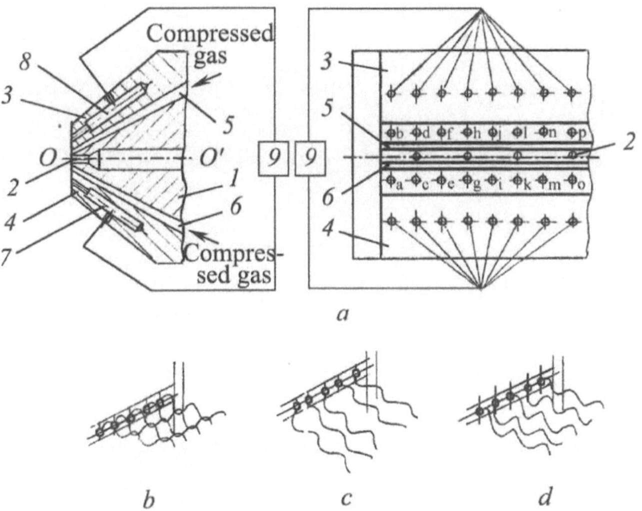

Fig. 3.13. An extrusion head that creates pulse gas flows: (**a**) general view; (**b**)–(**d**), sprayed fiber trajectories. *1*, housing; *2*, spinneret holes; *3* and *4*, cover plates; *5* and *6*, gaps; *7* and *8*, channels; *9*, gas distributor. Letters denotes channel outlets

When pulsating flows from hole pairs a-b, e-f, i-j, m-n... situated between holes *2* are switched on, then fiber trajectories will be in the horizontal plane (c) and look like sinusoids different in phase by 2π.

When the flow from holes c, h, k, p... and d, g, l, o... situated on the vertical axes of the holes are switched on alternately, fiber trajectories are of the type shown in Fig. 3.13d. Sinusoids are arranged in vertical planes passing spinneret hole axes and differ from the neighboring sinusoid value by π.

So, by using such a head [16], it has become possible to control the structure, density, and strength of melt–blown materials by alternating pulsating gas flows and by varying their velocity and treatment time.

Fiber modification by solid particles using special spray heads is commonly included in the melt blowing procedure.

The head described in [17] sprays three flat gas-dispersed flows simultaneously, namely, gas–powder with a modifier and two fiber-carrying flows above and below the first. This is attained by a torpedo-like splitter connected to a dividing plate *6* in a conventional multihole head (Fig. 3.14a). The splitter contains chamber *10* connected to a gas–powder generator which processes the solid dispersed modifier for polymer fibers. The chamber comes out onto the mouthpiece face as slot *12*. Spinneret holes are situated on both sides of the slot so that their axes together with the slot axes form angles α that converge in the spray direction.

Upon extruder actuation, the polymer melt circumvents torpedo *5* and splits into two parts. Polymer flows pass channels *7* and *7'* and are squeezed through the spinnerets as fibers. After this, generator *11* is switched on and compressed gas is fed into clearances *14* and *14'*. Under the joint action of gas flows *18*, *18'* and gas–powder flow *16*, the fibers are stretched, dispersed, and transported to the substrate by flows *17* and *17'* (Fig. 3.14b). Behind the intersection of the spinneret axes arranged at angle α to the spray direction, all of the flows join together. Solid particles stick to fibers while contacting them.

A difficulty often arises in manufacturing melt-blown materials of implementing local modification of fibers by solid particles of different natures with independent control of the distribution of each component across the material bulk. This can be resolved by the special extrusion head [18] shown in Fig. 3.15.

The head peculiarity is in the following. Fibers squeezed from a usual multichannel spinneret are sprayed by gas–powder flows prepared in generators *10–12* on a base of components of different natures, e.g., complexing agents with heavy metal ions, adsorbents, electret or magnetosolid substances, biocides, and surfactants. By varying generator regimes, the concentration gradient of each modifier in the final material can be controlled independently. This provides the basis for creating unique melt-blown articles with reduced

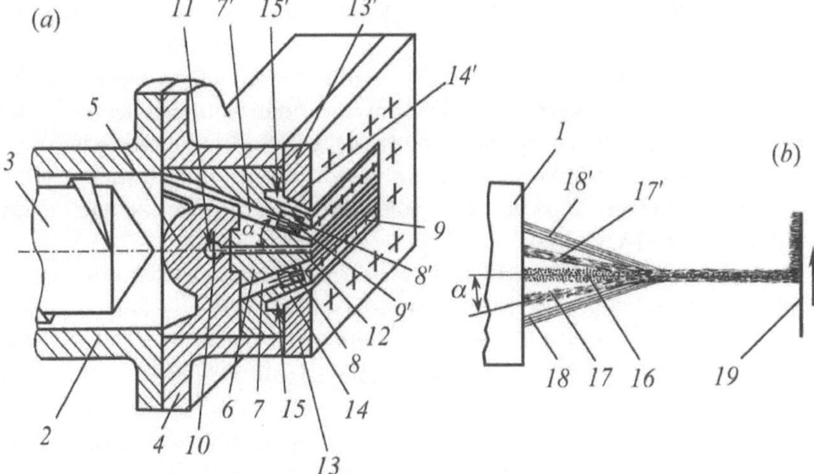

Fig. 3.14. An extrusion head simultaneously spraying fiber flows and their modifier: (**a**) principle view; (**b**) scheme of flows. *1*, mouthpiece; *2* and *3*, cylinder and extruder worm; *4*, housing; *5*, torpedo; *6*, dividing plate; *7* and *7'*, channels; *8* and *8'*, spinnerets; *9* and *9'*, spinneret holes; *10*, chamber; *11*, generator of gas–powder mixture; *12*, slot; *13* and *13'*, cover plates; *14* and *14'*, clearances; *15*, compressed gas manifold; *16*, gas–powder flow of modifier; *17* and *17'*, fiber flows; *18* and *18'*, gas flows; *19*, substrate

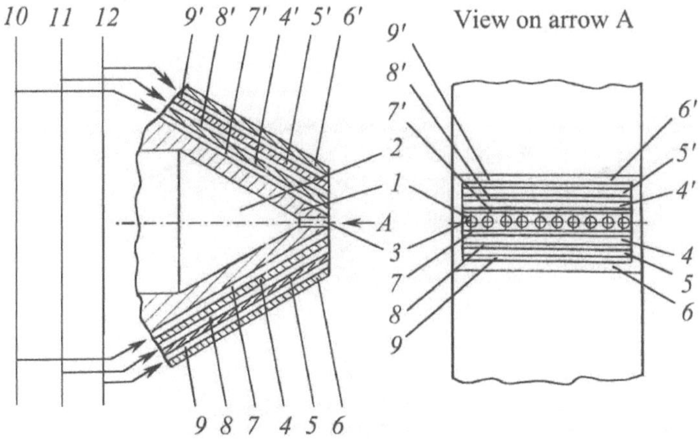

Fig. 3.15. An extrusion head for independently modifying melt-blown materials by different formulations of powder components. *1*, housing; *2*, cavity; *3*, spinneret holes; *4* and *4'*, *5* and *5'*, *6* and *6'*, cover plates; *7* and *7'*, *8* and *8'*, *9* and *9'*, clearances; *10*, *11*, and *12*, generators of gas–powder mixtures

modifier consumption without increasing the cost of the process or equipment complexity.

Electrical polarization of fibers can be realized with special spray heads. Such a head [19] polarizes fibers as the polymer melt squeezes through the spinneret without applying any outer electrical source. In contact with electrically closed plates of unlike metals, polymers polarize and form so-called metal–polymer electrets [20]. This principle is the basis for the extrusion head design illustrated in Fig. 3.16.

The steel housing of the head has cavity *2* connected with the extruder working volume and conjoined with cover plates installed with gaps *4* on the housing. The gaps are connected with a compressed gas source. The spinneret consisting of copper and aluminum parts separated by dielectric pad *7*, e.g., mica, is pressed into the housing. The spinneret press fit allows for electrical contact between parts *5* and *6*. Spinneret holes *8* are located so that their axes are inside the cross-sectional contour of pad *7* and their diameter exceeds the thickness of the pads, $d > h$.

The head performs a traditional melt blowing process except for one thing. While running through the spinneret, the polymer mass is brought into contact with parts *5* and *6* made of unlike metals. Because of a difference in

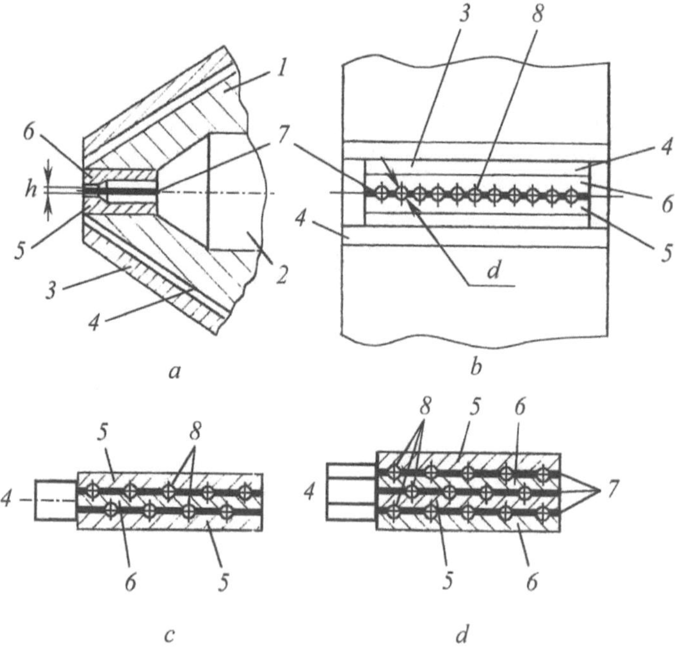

Fig. 316a–d. A polarizing extrusion head. *1*, housing; *2*, cavity; *3*, cover plate; *4*, clearance; *5* and *6*, copper and aluminum parts of the spinneret; *7*, dielectric pad; *8*, spinneret holes

the electrode potentials of Cu–Al and the melt conductivity, current flows in a closed circuit consisting of part *5*, polymer melt, and part *6*, housing. This causes electrical polarization of fibers, which is intensified during their extension and dispersion. The surface charge density of articles reaches 10^{-8}–10^{-5} C/cm^2.

Schemes of spinnerets are shown in Fig. 3.16c,d, where holes *8* are located in two or three rows and metal parts are separated by two and three dielectric pads, respectively, (oxide ceramics, filled teflon, textolite, etc.). Each pair of parts is closed to an outer circuit, which is the head body.

The operation of such a head does not necessitate any power source. Polarization is naturally combined here with the fiber forming processes. The head is characterized by simple design and maintenance.

Original spray heads as a rule blow special-purpose articles. The head described in [21] is intended for molding polymer packaging, sound- and heatproof materials incorporating film layers with a fixed fibrous mass. Its modification is presented in Fig. 3.17a. The design includes body *1* with two annular clearances *2* and *3* connected through channels *4* and *5* with working volumes *6* and *7* of the extruder. Along the circumference of an annular site between slots *2* and *3*, holes *8* are connected with extruder working volume *10* through channels *9*. Annular nozzle *11* is installed near holes *8* at an angle $15° < \alpha < 45°$ to them and is connected through channels *12* with a compressed gas source. On the site between slots *2* and *3*, hole *13* is connected to suction pump *15* through channel *14*. In the center of the annular site bound by slot *3*, hole *16* is connected through channel *17* with a compressed gas source. Gas tubes with regulating valves *18*, *19*, and *20* are connected with gas channels *12*, *14*, and *17*.

Operation of the head includes the following steps. Upon actuation of extruders *6* and *7*, films *21* and *22* are squeezed from slots *2* and *3*. They are inserted into the receiver of a hose aggregate. When valve *20* is open, compressed gas is fed through channel *17* and hole *16* into the hose cavity formed by film *22*. Gas blows the hose out and brings films *21* and *22* into contact below the solidification line *AB*. Polymer fibers are forced through spinneret holes *8* upon extruder *10* actuation. As soon as valve *18* is open, compressed gas starts to run into channels *12* and annular nozzle *11*. The gas flow picks up molten fibers, disperses them as a gas-polymer mixture *21*, and transports them into the clearance between films *21* and *22*. Films and fibers in a viscous-flow state are brought into contact and, upon cooling form, adhesive bonds between the material components above line *AB*. The pressure in the cavity between the films is regulated by valve *19* by evacuating excess gas through hole *13* and channel *14* using pump *15*. A layer of fibrous mass deposited between the films upon blowing the hose is compacted into layer *24*. Thus, a sheet material is obtained that consists of polymer films *21* and *22* adhesively bonded with fibrous-porous layer *24*.

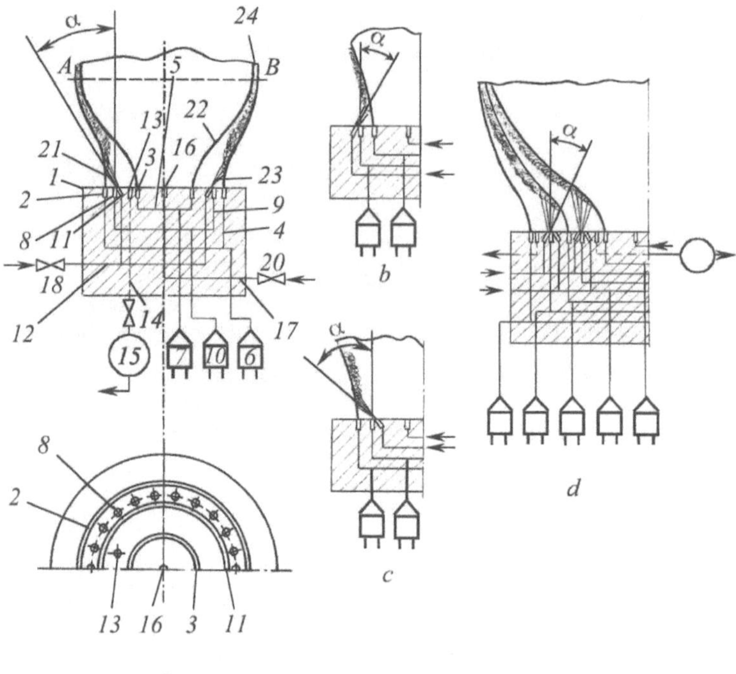

Fig. 3.17. An extrusion head for blowing fibrous-film materials: (a) three-layer; (b) and (c) two-layer; (d) five-layer materials

The heads illustrated in Fig. 3.17b and c are for polymer film production with an adhesively fixed fibrous-porous layer on either the outer (b) or inner (c) side of the hose. In Fig. 3.17d a head is shown for forming a five-layer material with a P-F-P-F-P structure, where P is a polymer film and F is a fibrous-porous polymer layer. The head differs from that shown in Fig. 3.17a by the number of channels for polymer melt and compressed gas. Its operating principle and actuation scheme are analogous to those cited before.

The head presented in [21] widens the technological potentialities of the extrusion methods for molding sheet and film materials on a polymer base. It can serve to control a wide range of deformation, soundproofing, thermophysical, and other characteristics of materials produced without changing their composition but by varying the technological regimes of the melt blowing process. The head allows increasing adhesion between the hose layers due to

- viscosity reduction of the melt sprayed on the film;
- pressure augmenting of the gas-polymer flow on the film surface;
- temperature increase in the film-fiber contact by overheating fiber at an optimum hose temperature;
- regulating the time when the contacting materials are in a viscous-flow state by changing the position of AB line of the hose solidification.

A problem often arises in creating melt-blown products of filling pipes, housings, and cavities where a fibrous mass imitates a body of revolution. This is advisable to fulfill at an elevated filler deposition velocity per cavity unit area and provide the possibility of fiber weaving.

An extrusion head [22] has been developed which can solve the problem. It is distinguished by the compact location of spinneret holes in conformity with nozzles through which spray gas flows.

The head is shown in Fig. 3.18a–c. Housing *1* has cavity *2* connected with the extruder working volume. Rotor *3* is movably joined with the housing by annular jut *4* inside a corresponding groove on the housing face. The cylindrical surfaces of rotor *3* contacting housing 1 have antifrictional coatings *5* serving as a sealant. The rotor is provided with a cruciform slit *6* connecting cavity *2* with spinneret channels *7*. The latter are in lid *8* fixed to the rotor. The rotor is installed rotationally on the housing through threaded bushing *9*. Inside the bushing is pipe *10* connected with a compressed gas manifold. Annular recess *11* on the cylindrical surface of the rotor connects with grooves *12* in lid *8* on both sides of each row of spinneret holes. Tube *13* connected to an aerosol generator is installed along the axis of cavity *2* in housing 1. It is joined movably to lid *8* by a sealing coating *5*.

The head operates as follows. Upon extruder actuation, the polymer melt is forced into cavity *2*, then flows into the cruciform slit in the rotor, and is squeezed through spinneret channels *7*. The gas flow is fed from the manifold through pipe *10* into annular recess *11* and issues from grooves *12*. It takes up molten fibers and extends and transports them to the substrate. An aerosol supplied through pipe *13* contains either solid (activated carbon, ferrites, polymers, etc.) or liquid-phase (solvents, plasticizers, corrosion inhibitors, etc.) modifiers. They are transported in the central part of the gas-polymer flow issuing from the head with minimum losses.

Thanks to antifrictional sealing coatings *5* on rotor *3* and pipe *15*, the rotor revolution (drive is not shown) about the body does not lead to any noticeable frictional forces or leaks of the melt and gas. When the rotor starts revolving, aerosol flow is surrounded by interweaving melt fibers, which reduces the probability of modifier losses still more. Cylindrical gas-polymer flows were found convenient for filling cavities in bearing members.

The head illustrated in Fig. 3.18d makes fibers for deposition at high speed (up to 10 g/s·cm^2), limited only by the extruder efficiency. The lines are geometrical centers of spinneret holes *7*, look like sinusoids originating from the center of the head symmetry, and increase with the distance from the center amplitude. Grooves *12* connected with a compressed gas manifold are equidistant to those lines on both sides of the holes. The head (d) operates similarly to the previous one (a–c). During operation, a concentrated gas-polymer flow is created whose density is constant in the radial direction.

It is evident that extrusion heads for melt blowing processes are an individual and intensely developing type of technological equipment. Their

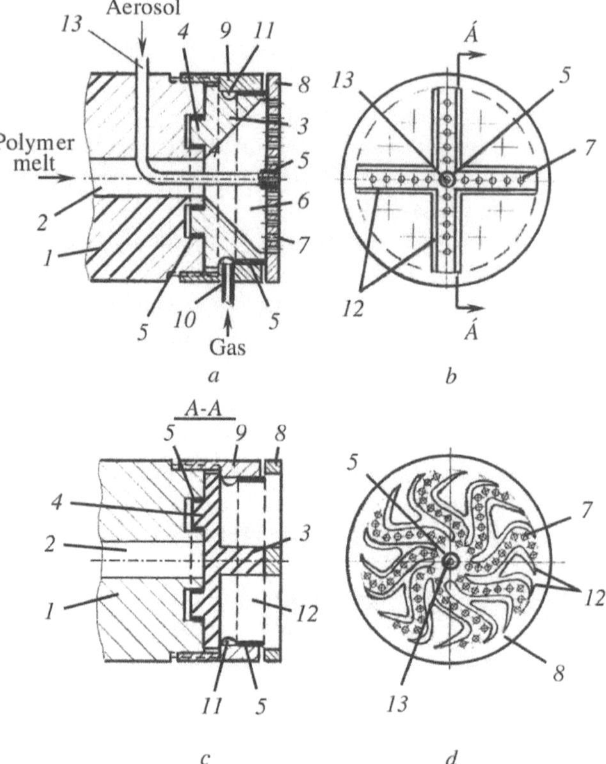

Fig. 3.18. An extrusion head for filling cavities in body parts with a fibrous mass: (**a**)–(**c**) side view, front view, and A–A section; (**d**) face of the head with elevated productivity

specific design, expanding fields of application, strict requirements for precision parts, and high cost put them among traditional equipment for plastics processing.

3.2 Auxiliary Equipment

The schematic diagram of an installation for a melt blowing process in Fig. 2.1 gives only a general idea of its technological instrumentation. Modern installations are more intricate and contain many pieces units of auxiliary equipment. In Fig. 3.19, the scheme of a melt blowing system produced by Reifenhäuser Co. is presented.

The system consists of five main blocks. Each block presents a separate unit of technological equipment, and a complex of standby machines and instruments is necessary to handle them. As an example, a complex is shown schematically in Fig. 3.20. The complex developed at Biax–Fiberfilm Corp.

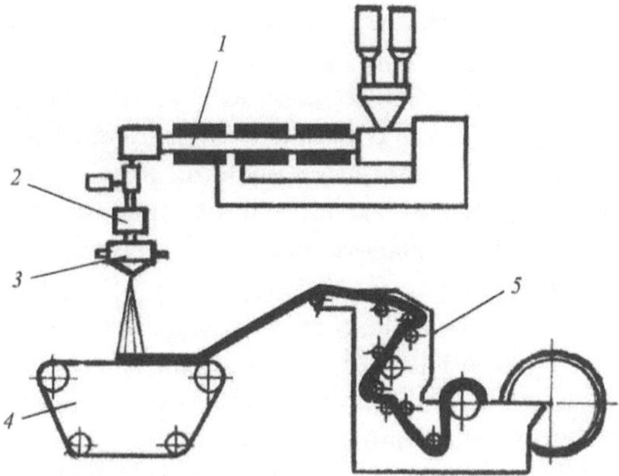

Fig. 3.19. The scheme of a Reicofil system for manufacturing melt-blown thermoplastic sheets. *1*, extruder; *2*, melt distributor; *3*, head with a spinneret; *4*, collector block; *5*, block for sheet cutting and winding

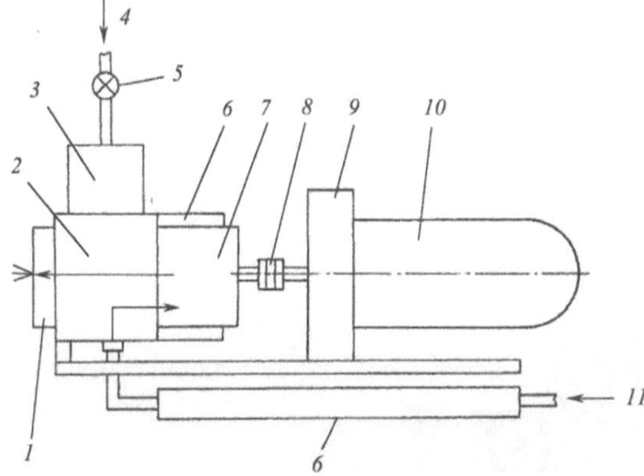

Fig. 3.20. The scheme of a melt blowing system developed by Biax–Fiberfilm Corp. [23]. *1*, spinneret; *2*, head block; *3*, air distributor; *4*, air flow from heat exchanger; *5*, valve; *6*, heater; *7*, pump for polymer melt; *8*, coupling; *9*, reducer; *10*, dc electric motor; *11*, polymer melt flow from extruder

(USA) provides for operation of blocks *2* and *3* (melt distributor and spray head, respectively) in combination with multichannel spinneret heads [13,14] with up to 80 spinneret holes per centimeter arranged in 10 rows.

The system also includes one or more gear pumps for even distribution of the polymer mass along the spinneret. Each pump is actuated by an individual drive for flexible control of melt flows. Only hot air is used for heating to exclude electrical heaters. The block has a modular structure, which permits fast replacement of sections for cleaning or maintenance [23].

Batch production of melt-blown products of a certain type necessitates unconventional design of auxiliary equipment. An example is a system for manufacturing tubular filtering elements [24]. It includes (Fig. 3.21) a cylindrical rotor along whose axis a spray head is installed connected to the extruder. Around it are forming mandrels rotated by a special drive.

The edges of the rotor are sealed by lids with hatches to remove finished products and connect the rotor cavity with the ventilation system. Operation of the system includes the following steps.

First, the generator of the melt-blown fibers is actuated, then the rotor and mandrels start to rotate. The fibers deposit on the mandrels as tubular articles whose shape is determined by the mandrel configuration. Rotor rotation intensifies fiber dispersion, which is important in processing materials with low melt index, e.g., secondary polymers. To reduce waste, the mandrels are arranged at $2\delta + D > H > 2D$ pitch, where D is mandrel diameter and δ is the article wall thickness. The sealed rotor isolates the working volume from the environment and makes it possible to remove toxic products of polymer breakdown by exhaust ventilation.

Application of electrical fields adds specific features to auxiliary equipment design. These include special polarization units, electrical insulation systems and original control systems, of the melt blowing process using electrical potentials.

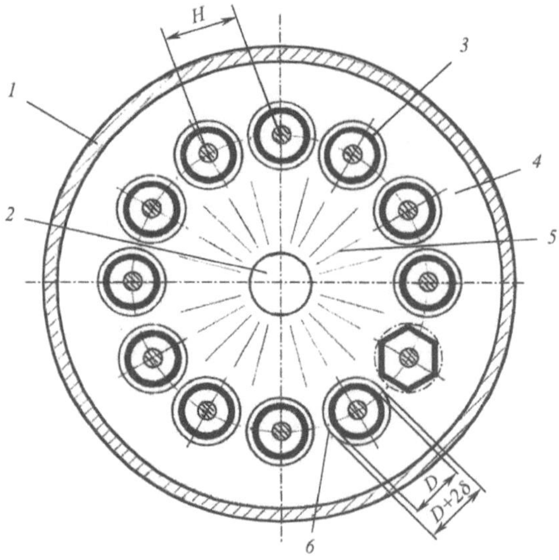

Fig. 3.21. A system for manufacturing tubular melt-blown articles. *1*, rotor; *2*, spray head; *3*, mandrel; *4*, lid; *5*, fiber flow; *6* and *7*, cylindrical and polyhedral articles

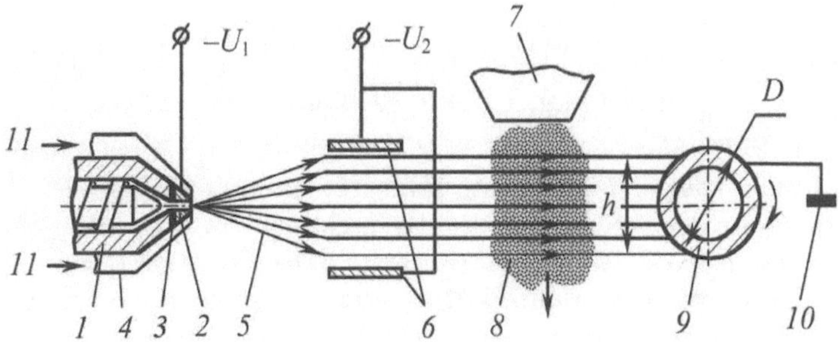

Fig. 3.22. An electrostatic device to produce fibrous articles. *1*, extruder; *2*, spinneret; *3*, dielectric pad; *4*, spray head body; *5*, gas-polymer flow; *6*, electrodes; *7*, aerosol generator; *8*, aerosol cloud; *9*, mandrel; *10*, ground grid; *11* gas flow

A device [25] for producing fibrous structural elements includes electrical equipment, which helps to reduce polymer waste and regulates the structure and density of fibrous materials. The scheme of the device is presented in Fig. 3.22.

The device includes spinneret *2* insulated from extruder *1* by dielectric pad *3* and connected to the negative pole of high-voltage source U_1. Between the extruder and forming mandrel *9*, there are plate electrodes *6* connected to the negative terminal of the U_2 potential source. Generator *7* of aerosol *8* in a liquid-drop phase is installed before forming mandrel *9*.

Polymer melt is squeezed through the spinneret connected to the U_1 source pole, which induces contact polarization of fibers. The fibers are homocharged and acquire negative polarity. The electromagnetic field of fiber flow *5* interacts with the permanent electrical field of electrodes *6* that has a negative potential. The force of fields' interaction moves outer layers of the flow to its axis and transforms the flow from divergent into plane flow. Under optimum regimes the width of flow *5* does not exceed diameter D of mandrel *9*.

Flow *5* passing through aerosol cloud *8* is accompanied by neutralization of the fiber polarizing charge when the dispersed aerosol is a conducting liquid. The degree of charge relaxation depends on the aerosol density, cloud size, flow velocity and charge state. When a dielectric forms an aerosol, it can affect only the fiber temperature. Depending on the polarity of the charged aerosol drops, their repulsion or adsorption on the fibers dominates during interaction between flow *5* and cloud *8*.

Grounding of mandrel *9* leads to fiber surface charge leakage and to denser fiber packing on the mandrel. Electrostatic attraction of charged fibers to a grounded mandrel is the reason for reduced polymer wastes because fewer fibers fly past the mandrel in the gas-polymer flow.

The system presented in [26] is intended for nonwoven fabric manufacture from polymer solutions, i.e., it does not use the melt blowing technique. Nevertheless, except for polymer preparation for processing and fiber solidification, the procedures are very close to those of the melt blowing process. In our opinion, an analysis of the system design is rather important for optimizing auxiliary electrical equipment of melt blowing systems.

The structure of the installation [26] is shown in Fig. 3.23. Polymer solution is forced from a special device 1 into spinneret 2 and flows from hole 3. Owing to the pressure drop due to flow and the high temperature of the solution, a flow of vapor is formed in the system cavity in the direction (shown by arrows) of the polymer mass motion. The flow is intensified by the revolution of baffle 4, which changes the direction of solution stream, disperses it, and forms fibrous mass 5.

The fibrous mass passes between parts 6, 7, and 8 which direct both vapor flow and the sheet to the substrate. Needle electrodes 9 in part 6 are connected to high-voltage source 10. A corona discharge is generated between needles 9 and circular target 7 grounded on a brush and commutative ring. The target can revolve independently of baffle 4. When revolving, its surface is cleaned by brush 11 to make the corona discharge stable.

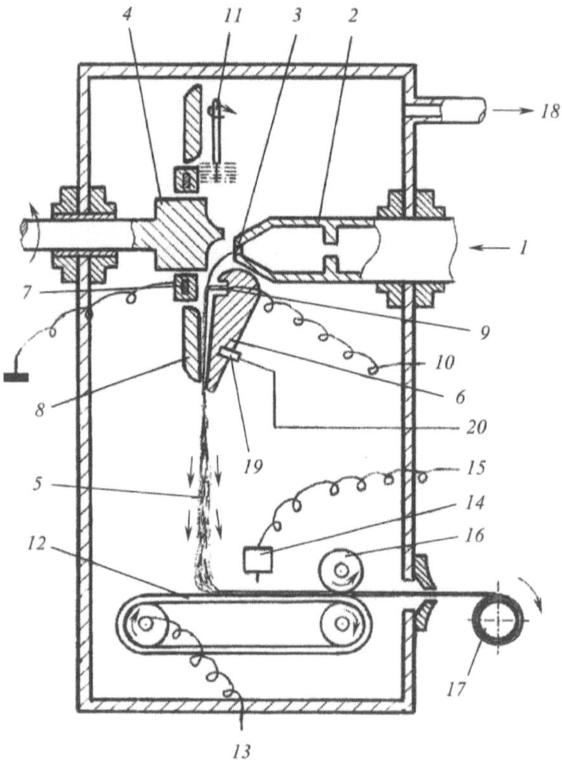

Fig. 3.23. A system for manufacturing nonwoven sheets from polymer fibers [26]

A fibrous mass is deposited as a fabric on the substrate formed as a conducting continuous band *12* set in motion by rollers. It can be either grounded or connected to constant voltage source *13*. Because it has opposite polarity, the fabric is attracted to band *12* and sticks to it. Forces of electrostatic attraction are, as a rule, sufficient to overcome the destructive effect of the turbulent vapor stream. If needed, an additional generator *15* is used. Then, the fabric is prepressed by roller *16* and wound on spool *17*. Unit *18* for solution regeneration improves the ecological parameters of the system.

The electrostatic attraction of fibrous fabric to the substrate is a major factor influencing the final product quality. To govern the degree of fibrous mass charging, meter *19* is installed in part *6*. The corona discharged and oscillating under a baffle effect mass induces an alternating current in the meter proportional to the charge value of the mass. The signal is picked up by recorder *20*, which warns the operator of insufficient charging of the mass to avoid the risk of rejects.

The specifics of the melt blowing technique involving unusual procedures lead to the development of *new designs* of auxiliary equipment.

A typical example is a device for molding tubular melt-blown elements [27]. Its design includes moving joints where the forming mandrel rotates and reciprocates about the spray head. As soon as the sprayed tubular element has reached a preset thickness, it is automatically moved to a certain pitch at the mandrel edge, and the formation of the next portion of continuous tube is formed on the cleared surface of the mandrel. The distinguishing feature of the design is in the velocity of tubular element reciprocation relative to the immovable spray head, which is preserved constant independently of whether or not the finished article is moved away from the rotating mandrel. This happens because the finished tube is moved away in a direction opposite that of the mandrel, which has the sprayed tube on it, at a velocity twice as high as that of the mandrel linear motion. The device is helpful in molding continuous melt-blown tubes with a uniform thickness, which can be controlled in a wide range.

The trend to attain a high-efficiency melt blowing process is often in contradiction with the requirement to deposit fibers accurately onto the local substrate spots with a controlled degree of cohesion. Biax–Fiberfilm Corp. has created an unusual setup [28], which forms articles from cohesively unbonded and practically continuous fibers subjected to a high degree of extension. Interwoven and highly oriented disconnected fibers are processed into filtering elements that display higher productivity compared to analogous products made of fibers welded by bridges. With this aim, the property of oriented fibers to shrink at a temperature higher than that of the polymer glass transition, which lead to a 20-fold increase in density of melt-blown materials from 0.01 to 0.20 g/cm^3 is used.

The setup (Fig. 3.2) consists of spray head *1* connected through heated pipe *2* to extruder *3*. Melt-blown fibers *4* are supplied from the head into a

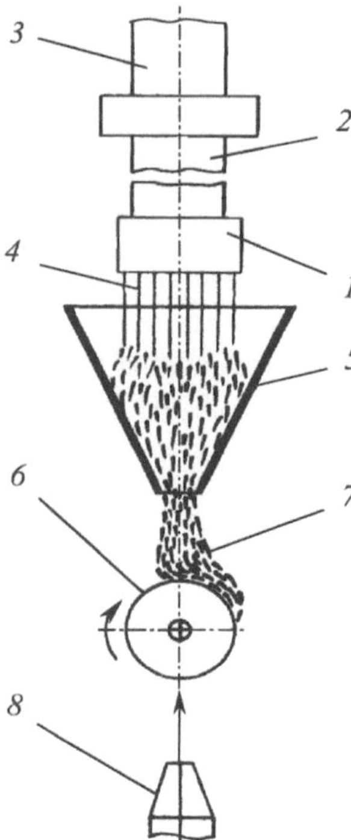

Fig. 3.24. A system for molding articles by orienting and shrinking melt-blown fibers

pyramidal chute *5* which directs the fibers to revolving ($n = 10$–250 rpm) mandrel *6*. Mass *7* of strongly entangled incoherent fibers is wound on the mandrel and is blown underneath by a flow of hot air ($T = 70$–$265°C$). It issues from blast device *8* at an initial speed $\nu = 150$–300 m/s. The air heats the fibrous mass, which shrinks on the mandrel and forms a tubular article. Analogous equipment [28] is used to produce fibrous materials intended for sleeping bags, gloves, winter jackets, pullovers, etc.

Often there is a problem of biaxially extending of fibrous fabrics for isotropic reinforcement. It was attempted to solve the problem by traditional methods, but tearing of separate fibers and breaking welding bridges between them and frequently occurred, consequently, the fabric narrowed. A device [29] for biaxially extending of fabrics formed by unoriented fibers has overcome this drawback. It involves two sets of drawing rollers grooved on a generatrix. In the first set, the grooves are parallel; in the second, they are perpendicular to the roller axes. The groove profile is sinusoidal with $H < 1.0$ mm pitch at a specific mass of fabric up to 10 g/m^2.

When the fabric passes the first pair of rollers whose grooves are parallel to the roller axes, the fabric is extended longitudinally. Sinusoidal roller grooves engage with juts on another roller as in a pair of pinions. The fibrous mass in the clearance between them is drawn to a value equal to the length of a

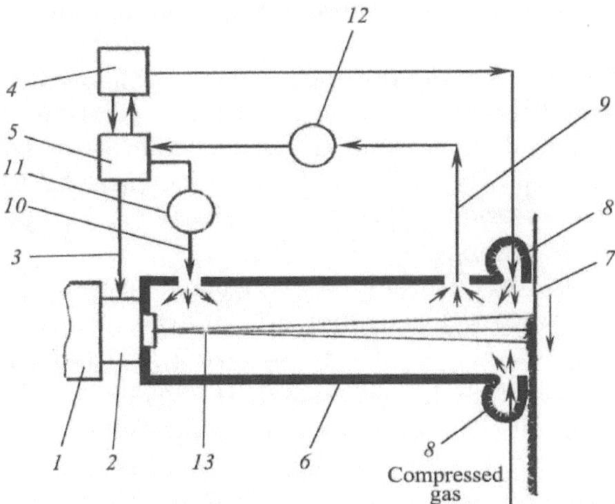

Fig. 3.25. A device [30] for manufacturing fibrous nonwoven materials

sinusoidal period. Pressure rollers avert fabric slippage. Using the previously
mentioned parameters for the grooves and fabric, neither fiber rupture nor
fabric narrowing were recorded during longitudinal extension.

After this, the fabric undergoes transverse extension by passing over
stretching fixtures and rolling by rollers whose are grooves perpendicular to
the roller axes. The finished fabric is reeled on a spool. These procedures are
repeated to reach the needed extension value and homogeneity of properties.
Thus, fibrous fabric the setup described processed in [29] becomes softer than
the initial material and acquires a decorative structure.

Modification of polymer fibers by substances that can be in any aggregate
state creates the possibility of their physicochemical interaction with the
polymer melt. Because this might impair service conditions, the reaction
volume of the melt blowing setups is made airtight and intended for spray
gaseous dispersed systems circulating in closed containers. An example of
such a setup [30] is shown in Fig. 3.25.

Extruder *1* is equipped with spray head *2* connected through channel *3*
with compressed gas source *4* and aerosol generator *5*. Hollow casing *6* is
hermetically fixed on the extruder, and conveyor belt *7* on its face is super-
imposed. Discharge chambers *8* are installed on the casing face and connected
to a compressed gas source. The cavity of casing *6* is connected to aerosol
generator *5* through channels *9* and *10* where delivery *11* and suction *12*
pumps are placed.

Upon extruder actuation, molten polymer fibers are squeezed from head
2. A gas flow from source *4* is fed to generator *5*, and aerosol is supplied
along channel *3* to the forcing zone. Fibers are picked up by the flow and
are deposited on belt *7* as a jet of aerosol-polymer mixture *13*. When the

conveyor is switched on, a continuous layer of fibrous mass is formed on the belt.

The fibers are modified in contact with the dispersed phase of the aerosol. To prolong the contact, it is worthwhile to feed the aerosol from generator *5* to cavity *6* along channel *10* using delivery pump *11*. Used aerosol is returned by pump *12* through channel *9* into the generator. Compressed gas is fed to the discharge chambers to reduce modifier leakage from the casing. The gas excess is evacuated from the casing by pump *12*.

The fibers can be modified by toxic, explosive, or inflammable agents using a device described in [30]. A closed gas circulation system makes the melt blowing process reliable and ecologically safe in operation with any modifiers.

The melt-spinning process is the nearest analogue to the melt blowing technique. The difference is in the method of fiber extension which takes place under its own weight, while gas streams participate in controlling the processes of fiber cooling, interweaving, and packing on the substrate [31]. An analysis of auxiliary devices for the spinning process could also show ways of refining the melt blowing process.

In the late 1980s, the Reifenhäuser Co. patented a set of systems for producing fibrous nonwoven materials. A typical design of such a system [32] is presented in Fig. 3.26.

Polymer fibers *2* are extruded through multichannel head *1* and then blown off by airflow in cooling compartment *3*. The airflow from nozzles *4* of chambers *5* and *6* of intensive and additional cooling is connected with a compressed gas manifold. The flows make fibrous roving narrower like a wedge directed to slot zone *7* where it is drawn through the clearance between screens *8* swiveling about their horizontal axes. Then the fibers pass

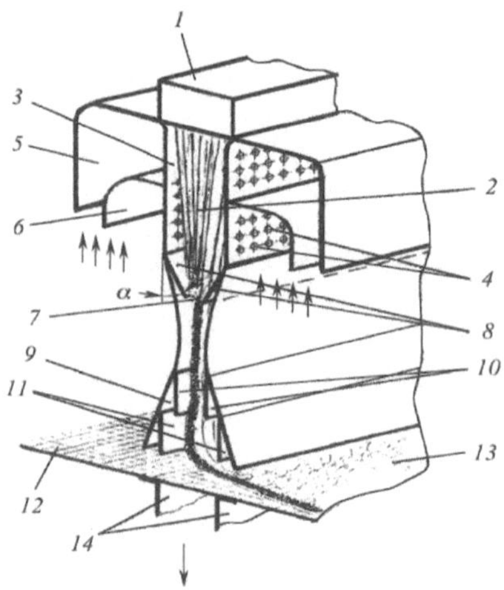

Fig. **3.26.** Reifenhäuser's system for manufacturing fibrous nonwoven materials

diffusion compartment *9* fit with swivel screens *10* and blades *11*. After this, the roving is laid by its loops on air permeable conveyor belt *12* where fibrous fabric *13* is formed. Beneath the conveyor belt, gates *14* are installed to permit displacement over the belt and shaping into a chute for exhaust air evacuation.

By varying turn angle α of screens *8*, *10* and blades *11* relative to the extrusion direction, the fiber packing pitch on the conveyor belt and the fibrous fabric density can be changed. When the pitch is larger then the preset pitch, the deflection of blades *11* from the vertical line is increased. When the fabric density exceeds that required, screens *8* are turned so that angle α increases. The intensity of fabric cooling is controlled by varying the width of the conveyor belt sucked through the airflow by displacing gates *11*.

The simplicity and compactness of this structure furnishes wide possibilities in controlling the technological parameters, which have led to a decision to launch batch production of systems for license sales.

Nonwoven fibrous materials can be used in industries other than the textile industry. Historically, in the manufacture of high-strength, unshrinking filaments, fibers, strings, and ropes, preference has been given to artificial silk, then to nylon; in the 1970s, to polyesters, later to polyethylene terephthalate and since 1985, to new grades of high-strength polyesters. One of the modern trends in improving the strength and stability of continuous articles on the materials mentioned base is realized by the spinning process.

A typical structure of auxiliary equipment [33] for producing continuous fibrous articles consists of a 5–9 m long vertical hollow column with a 30-cm inner diameter. A bunch of fibers extruded through a spinneret is let down into the column, drawn under its own weight, after which it passes through four to seven heating zones and is woven in the lower part of the column into threads, braids, strings, etc. by air flows from a perforated cone installed about the column axis. Upon removing static electricity the product is reeled on spools. The productivity of the system developed at Hoechst Celanese Corp. at a fiber velocity of about 3000–5800 m/min is 5–25 kg/h. The product is used to manufacture cord, cloth for canvases, mats for road construction, drive belts, ropes, composite layered materials.

As seen, technological equipment for the melt blowing process has an unusual structure with modifications, which reflect the specifics of the method and determine the external appearance of the devices. This shows that melt blowing instrumentation has merged with standard industrial equipment for plastics processing widely applicable in engineering.

4. Structure of Melt-Blown Polymer Fibrous Materials (PFM)

The method of melt blowing differs essentially from traditional methods of plastics processing. It proceeds at the boundary of oxidizing destruction of polymers when their technological properties are specified by low viscosity of the polymer mass in a viscous-flow state. This brings about active oxidation of the fiber surface layer, and intensified adhesive interaction between fibers and solid modifiers. The indicated phenomena, together with adsorptive interaction of fibers with modifiers in a liquid or gaseous state, condition the specific structure of melt-blown materials. They are characterized by the fiber diameter, the density of cohesive bonds between fibers, the porosity of the fibrous mass, the modifier concentration, etc. The structural parameters are determined by the technological regimes of material manufacture.

4.1 Major Structural Parameters

Polymer fibrous materials (PFM) obtained by the melt blowing method are heterogeneous systems. They are composed of at least two phases (in the absence of modifiers and fillers), namely, a fibrous polymer matrix in the shape of fibers chaotically disposed in space and adhesively bonded in contacts, and a combination of vacancies between them.

Many of the PFM functional characteristics (in particular, for filtration) are defined by the parameters of the materials fibrous-porous structure. To these belong the fiber mean diameter ($\bar{d}_f$), the fiber distribution by diameters, the volume density (ρ) and general porosity (P) of the material, the pore section distribution by size, the fiber specific area (S_{sp}), and others.

The fluctuation of temperature and gasodynamic regimes of polymer melt spraying in PFM production creates the possibility of controlling these parameters (see Sect. 4.2.). Moreover, PFM structure can be anisotropic due to the peculiarities of the melt blowing technique, which requires special accuracy of the methods used to investigate the structural characteristics of such materials.

In [1, 2], a comparative estimate is given of different structural and analytical methods of melt-blown materials research by criteria for their labor intensity, rapidity, and reliability; the interrelations among the structural parameters of PFM are also described.

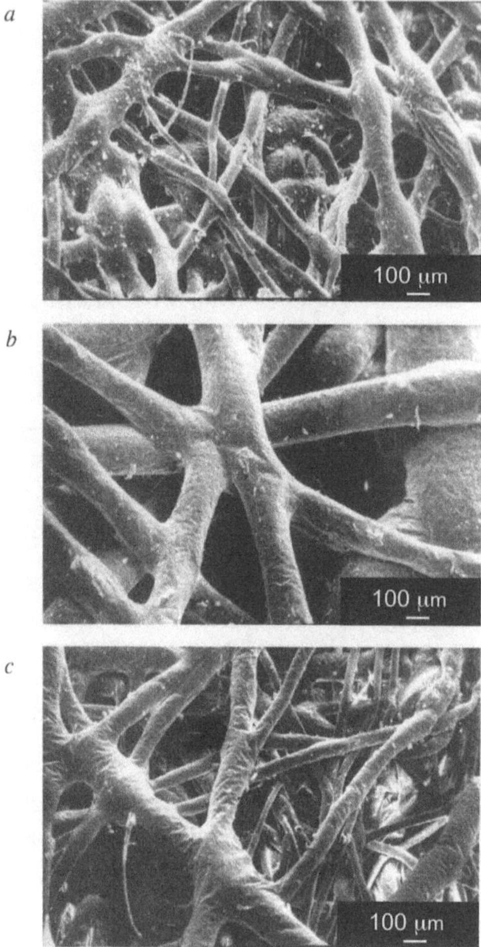

Fig. 4.1. Electron-microscopic images of melt-blown PE-based PFM: (**a**), (**b**), (**c**), samples corresponding to 1, 3, 5 in Table 4.1

The simplest PE-based PFM samples with different fibrous-porous structures were investigated (Fig. 4.1).

To determine the structural parameters of PFM, the following methods were chosen: measurement and weighing of the samples (ρ, P), optical and scanning electron (SEM) microscopy (P, $\bar{d}_\mathrm{f}$, etc.), and sorption analysis (S_sp).

The weight-space method of measuring the density and general porosity of materials provides sufficiently high accuracy; it is rather simple and is not time-consuming. General porosity is determined by the formula

$$P = (1 - \rho/\rho_p) \cdot 100\% ,\tag{4.1}$$

where ρ is the volume density of a PFM sample, and ρ_p is the polymer density.

The distribution of vacancy sections in a PFM matrix by size, their mean characteristic dimensions d_l (by length) and d_w (by width) [3], general porosity P, and fiber mean diameter $\bar{d}_\mathrm{f}$ were determined by automatic optical image analysis (AOIA) using the Min–Magescan analyzer. For this purpose sheet PFM samples were impregnated with epoxy resin (cold pouring) upon which setting cuts were made at different depths of the sample parallel to the sheet surface. Samples were then polished by diamond pastes of various graininess till a mirror-like smooth surface of the cut was obtained (Fig. 4.2). An image analyzer was assigned the program of constructing histograms of d_l, d_w and d_f distributions by size and calculating the mean parameters of $\bar{d}_\mathrm{l}$, $\bar{d}_\mathrm{w}$, $\bar{d}_\mathrm{f}$ [3, 4]. General porosity was calculated from the relation

$$P = (S_1/S_2) \cdot 100\% \, , \tag{4.2}$$

where S_1 is the area of resin-filled vacancy sections detected by the analyzer, and S_2 is the total area analyzed. As a rule, the porosity calculation using AOIA yields reduced values of P (Table 4.1).

On the contrary, when determining d_f parameters, the method overstates their values (Table 4.1) because the analyzer can not discern between images of a single fiber and twinned fibers accidentally occurring in the scanning path. Estimates of d_f (d_l, and d_w as well) presented by computer processing of SEM images are more correct [5].

It is worthwhile using an updated optical-microscopic method (OM) for constructing histograms of d_f distribution by size and $\bar{d}_\mathrm{f}$ calculations [1]. It accounts for PFM structural peculiarities, including the polydispersed composition of fibers and material structural anisotropy. The latter occurs because one of the sides of the PFM article contacts the forming substrate during manufacture. Due to intensive heat removal through the substrate

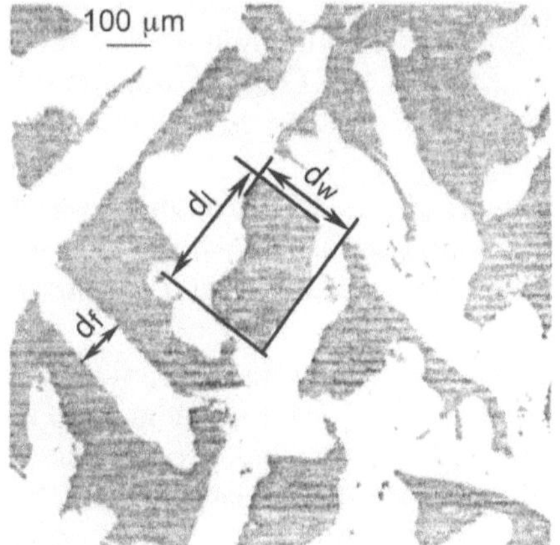

Fig. 4.2. Optical image of a longitudinal cut of sample 3 (Table 4.1) with parameters d_l, d_w and d_f

Table 4.1. Parameters of PE-based PFM sample fibrous structure

Sample No.	$\rho(\mathrm{kg/m^3})$	$P(\%)$		$\bar{d}_\mathrm{f}(\mu\mathrm{m})$			$\bar{d}_1(\mu\mathrm{m})$	
		Measurement and weighing[b]	AOIA[c]	OM[d]	AOIA	SEM	AOIA	SEM
1[a]	369±7	61.6	27.0	18±2		56±2	295±10	128±15
2	419±4	54.4	–	59±3	–	–	–	–
3	373±7	59.5	38.7	125±6	166±6	145±3	463±20	211±10
4	320±10	65.2	–	179±6	–	–	–	–
5	250±5	72.8	45.6	25 (estimate)	–	67±3	370±10	95±3

[a]Sample 1. HDPE; the rest. LDPE; [b]From (4.1); [c]From (4.2); [d]By the method in [1].

Sample No.	$\bar{d}_\mathrm{w}(\mu\mathrm{m})$		$\bar{d}_1/\bar{d}_\mathrm{w}$	$\bar{d}_1 \times \bar{d}_\mathrm{w}$ $(\mathrm{mm^2})$	$S_\mathrm{sp}(\mathrm{m^2/kg})$	
	AOIA	SEM	AOIA	AOIA	By V_m	By BET
1	192±7	59±4	1.54	0.057	92	109
2	–	–	–	–	82	–
3	253±13	116±10	1.83	0.117	51	42
4	–	–	–	–	37	40
5	240±10	48±2	1.54	0.089	–	–

during melt solidification, a fibrous matrix is formed with an irregular d_f distribution across the PFM thickness. Because of this, the relative number of d_f measurements on both the surface and inside the sample is preset by the technique. For this purpose, the material thickness, approximate mean diameter of fibers d_0, and relation $k = h/d_0$ must be found. Measurement ratios in the inner layers (N_1) and on the sample surfaces (N_2) should roughly correlate in the proportion $N_1/N_2 = k/2$.

The reliability of estimates is determined by the criterion: whether the real number of parameter measurements reaches the minimum needed number when statistical distributions are justifiable [6, 7]. Only an appreciable number of measurements produces results that reflect the real character of distribution.

It has been established that the sequential increment of the d_f measurement number from 50 to 800 leads to a maximum shift of the d_f distribution curve by size into the region of lower values of the parameter and to conversion of the differential curve $\varphi(d)$ pattern (Fig. 4.3).

Along with this, other characteristics of the distribution function also vary. It turned out that the minimum number of d_f measurements should be 600-800 to obtain reliable analytic results. It is seen that the right-hand portion of the decline of the differential curves $\varphi(d_\mathrm{f})$ under investigation slopes more gently then the left, and the peak is more acute in contrast to the curve of normal distribution. The scattering of d_f values compared to the mean (variation ratio), does not, in fact, depend on the number of

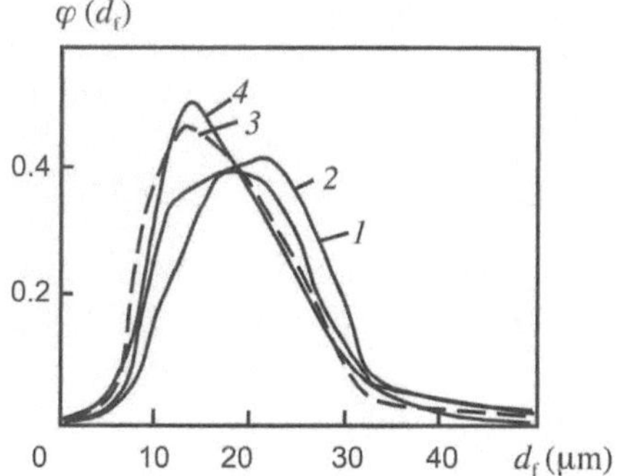

Fig. 4.3. Differential curves of fiber distribution by diameter for different numbers of diameter measurements: *1*, 200; *2*, 300; *3*, 600; *4*, 1000. PFM sample 1 from Table 4.1

measurements. Considering the established inconsistency between experimental data and normal, the logarithmic normal, and the Weybull–Gnedenko distribution, methods have been recommended in [1] for calculating the confidence interval of d_f using Chebyshev's inequality [6]. The method developed makes it possible to obtain reliable data reflecting the character of PFM fiber diameter distribution by size and mean $\bar{d}_f$ value.

An important structural and filtering characteristic of PFM is the vacancy size between fibers. As seen from Figs. 4.1 and 4.2, vacancy cross sections are anisotropic by shape and size. In this case, it is convenient to use the distribution of section size by length d_l (maximum Feret diameter) and width d_w (Fig. 4.4), as well as corresponding mean values $\bar{d}_l$ and $\bar{d}_w$, shape factor $\bar{d}_l/\bar{d}_w$ and flow section $\bar{d}_l \times \bar{d}_w$ (Table 4.1).

AOIA seems to be a fairly justifiable method for determining the size of PFM filtering channels in separate filtering planes. Along with this, another method based on computer processing of SEM images of PFM (Fig. 4.1) gives additional information on the size of flow sections and accounts for PFM specifics as volume (depth) filtering materials (see Chap. 6).

PFM consist of smooth nonporous fibers (Fig. 4.1). Their specific area is, apparently, not large. Because of that, the problem of defining S_{sp} can be best solved by the method of low-temperature adsorption of inert gases (nitrogen, krypton, argon, and others) [8].

The specific area was found by isotherms of low-temperature adsorption of nitrogen (Fig. 4.5) using the Brunauer-Emmet-Teller (BET) theory and the method of a characteristic point on the isotherm corresponding to a specific adsorption value (V_m, cm^3/g) for formation of an adsorbate monomolecular layer on the polymer fiber surface ($\rho = \rho_m$) [9]. The specific area (S_{sp}, m^2/g) was calculated by the formula

$$S_{sp} = V_m N_A S_0/\omega_m \, , \tag{4.3}$$

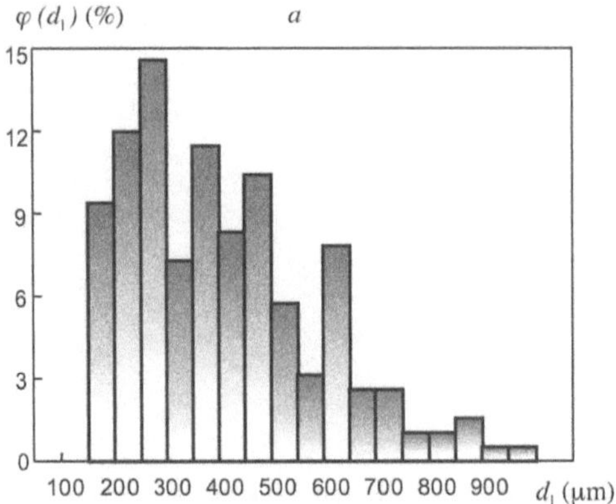

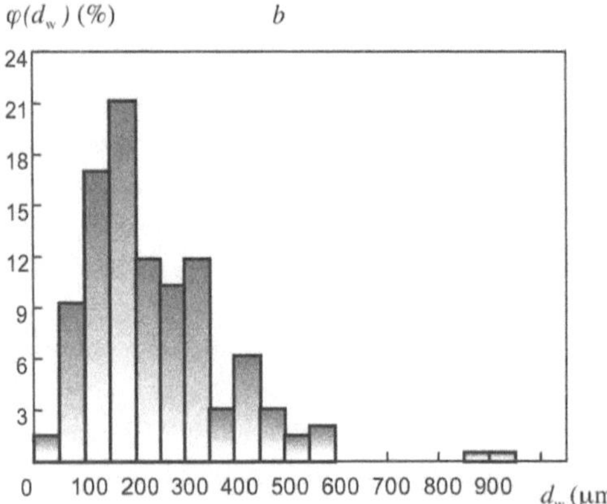

Fig. 4.4. Distribution histograms of PFM flow sections by size in two mutually perpendicular directions: (**a**), lengthwise (d_l); edgewise (d_w) (see sample 5, Table 4.1)

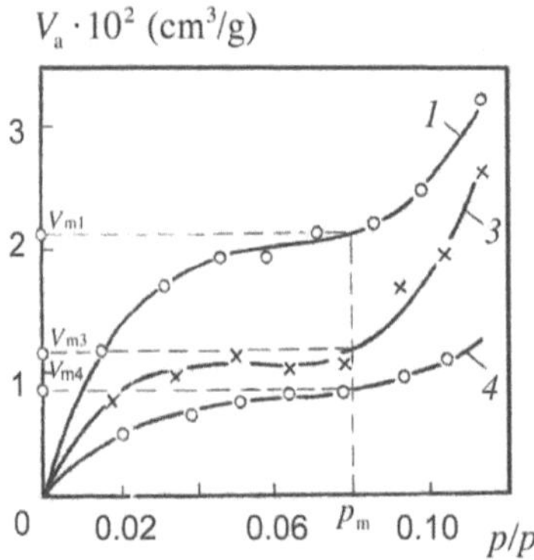

Fig. 4.5. Isotherms of low-temperature adsorption of nitrogen on PFM samples: V_A, liquid adsorbate volume; p/p_s, adsorbate relative pressure. Curves are indicated according to the numbers of samples in Table 4.1

where $N_A = 6.023 \cdot 10^{23}$ mole^{-1} is Avogadro constant, $S_0 = 16.2 \cdot 10^{-20}$ m^2 is the area occupied by the nitrogen molecule, and $\omega_m = 22.4 \cdot 10^3$ cm^3/mole is the molar gas volume.

When processing experimental data by the two methods (BET and the characteristic point), good enough correlation was obtained for S_{sp} values for each sample (Table 4.1).

The PFM specific area is determined by fiber diameter. Values of S_{sp} and d_f within the interval 15 to 180 µm are interrelated through a linear correlation (Fig. 4.6).

An analysis of the data shown in Table 4.1 of the fibrous-porous structure of PFM has drawn us to the following conclusions.

1. The density reduction of PFM at constant $\bar{d}_f$ (samples *1* and *5*) is accompanied by increasing general porosity and the size of the cross sections $\bar{d}_1 \times \bar{d}_w$ of vacancies between fibers.
2. With similar volume density and general porosity, materials with more fine fibers show fewer cross sections of vacancies between fibers (samples *1* and *3*).
3. The surfaces of fibers developed and the twisting of vacancies between them provide mechanisms for realizing depth filtration using PFM as a filtering material (see Sect. 6.2).
4. PFM with a considerable vacancy volume and developed fiber surface are promising materials for adsorptive and biotechnological systems (see Chap. 9 and 10).

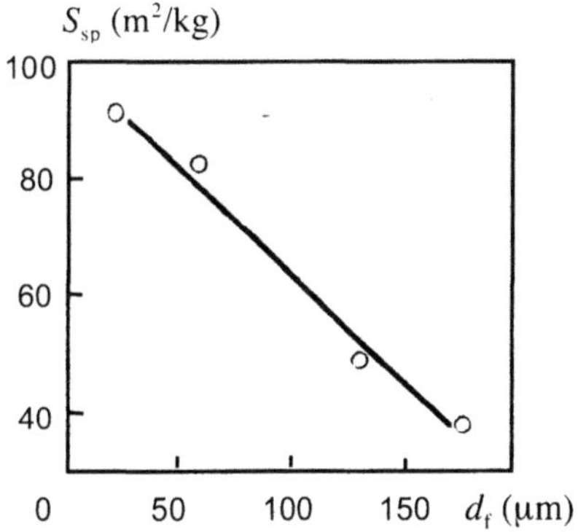

Fig. 4.6. PFM specific area dependence on mean fiber diameter

4.2 Effect of Different Technological Regimes on PFM Structure

The mean diameter $\bar{d}_f$ of fibers and the density of melt-blown materials are easily controlled structural parameters that predetermine the service characteristics of different purpose materials and articles made from them. The parameters depend on four independently controllable technological factors that govern the efficiency of melt-blowing process: rotational speed n of the extruder worm, pressure p, temperature T of the spray gas, and the distance L from the spray head to the substrate.

The experimental data cited following were obtained using a laboratory setup whose diagram is given in Fig. 2.1. The extruder with a worm diameter of 20 mm was fitted with a single-hole type I spray head (see Fig. 3.1a). Extrusion regimes are described in Table 4.2.

Table 4.2. Temperature in extruder zones and spray head for melt blowing polymer materials

Material	Temperature in the zones (°C)			
	1st	2nd	3rd	Head
LDPE	200	310	450	440
LDPE + SF[a]	190	320	440	430
PA	280	330	390	380
PA + SF	285	350	390	380

[a] SF, dispersed strontium ferrite (20 mass %)

The fiber diameter depends strongly on the technological parameters of the melt blowing process. It is seen from Fig. 4.7, that all other conditions being equal and averaged by varied parameters T, p, L (Figs. 4.8–4.10), increased rotational speed of the worm leads to the growth of the fiber diameter.

The greatest d_f growth is characteristic of polyethylene (curve *1*), and the flattest curve *3* reflects polyamide. Dependencies *2* and *4* corresponding to ferrite-filled materials practically coincide. This fact proves that the diameter of fibers from filled thermoplastics is determined mainly by the degree of filling.

Increased rotational speed of the worm leads to growing efficiency of the process (Fig. 4.7b); maximum values are shown by filled materials. It can be anticipated that the behavior of $d_f = f(n)$ dependencies is governed by the overlap of two interrelated factors, extrusion efficiency and extrudate viscosity. Increased productivity under other conditions of extrudate spraying are equal and lead to increased diameter of fibers. The degree of their extension upon spaying depends considerably on the polymer mass viscosity. PA melt

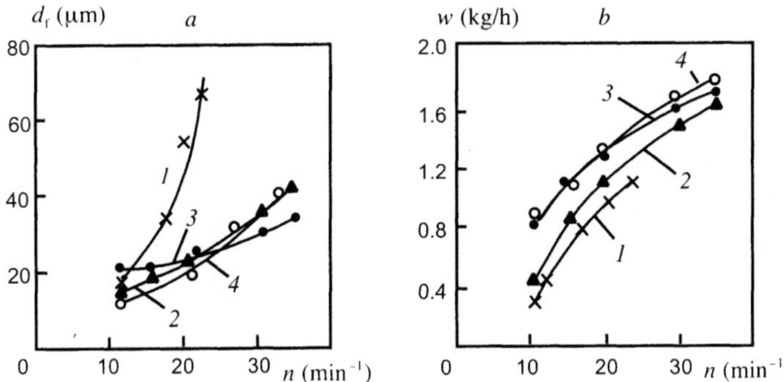

Fig. 4.7. Mean fiber diameter (a) and productivity (b) of polymer material melt blowing versus worm rotational speed. *1*, LDPE; *2*, LDPE + SF; *3*, PA; *4*, PA + SF

viscosity is the lowest, thanks to which curve *3* in Fig. 4.7a is the flattest. Filled material binders acquire high fluidity due to the additional heat returned to the polymer melt by filler particles. The latter have relatively low heat capacity and are heated to higher temperatures than the polymer during extrusion. Probably, one of the factors affecting the degree of filled fiber extension is the difference between the displacement velocities of the polymer binder and filler particles adhering to it inside a gaseous flow. That is why curves *2–4* are located below curve *1* that corresponds to a highly viscous melt of LDPE.

When the temperature of the spray gas rises, the mean fiber diameter of the materials studied exponentially approaches some minimum values (Fig. 4.8a). Curves *3* and *4* reflecting PA behavior are located above the LDPE curves (*1* and *2*). This is a confirmation of the heat-induced increase in the fluidity of PA-based extrudates within the temperature range studied, in contrast to LDPE-based materials whose melt fluidity rises insignificantly under these conditions. This supposition has been experimentally confirmed as will be shown below (see Fig. 5.1). Fibers from filled PA are thinner than those from unfilled PA owing to the larger heat quantity accumulated by the filled extrudate and the elevated PA fluidity that serves as a binder in the filled system. Curves *1* and *2* related to materials based on LDPE practically coincide. This means that the spray gas temperature in fiber formation from these materials is a more important factor than filling.

The mean fiber diameter dependence on spray gas pressure has the same character (Fig. 4.8b), that is, with an increase in pressure, the fiber diameter is reduced exponentially. Note that a pressure rise leads to cooling of the spray flow. If it were possible to maintain a constant temperature in the gas flowing under various pressures, then the exponential curves in Fig. 4.8b would occupy lower positions. All other conditions being equal, the fiber diameter of filled materials (curves *2* and *4*) is less compared to fibers of

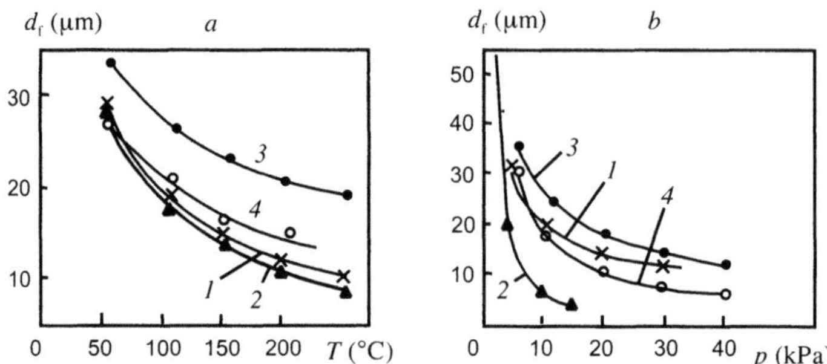

Fig. 4.8. Fiber diameter dependence on temperature (**a**) and pressure (**b**) of spray air. Curve designations are the same as in Fig. 4.7

unfilled materials (*1* and *3*) due to the previously mentioned spreading of binders when they absorb heat stored by filler particles.

The dependence of fiber diameter on the distance L between the spray head and the substrate to which the fibers are extended is a curve with a minimum (Fig. 4.9). First, with an increase in L, fiber extension and thinning conditions improve. Further, an increase in L intensifies molten fiber cooling and causes shrinkage and diameter growth. To a greater extent this is unequal to LDPE (*1* and *2*) and less so for PA (*3* and *4*) whose fibers do not, in fact, shrink under the L values investigated. The general trend here is the location of curves for filled materials that are lower than those of unfilled materials.

The density of melt-blown materials can be controlled by varying the technological parameters across a wide range (Fig. 4.10).

Density increases with growing rotational speed of the worm (Fig. 4.10a) due to increased fiber diameter. The density of fibrous materials made of

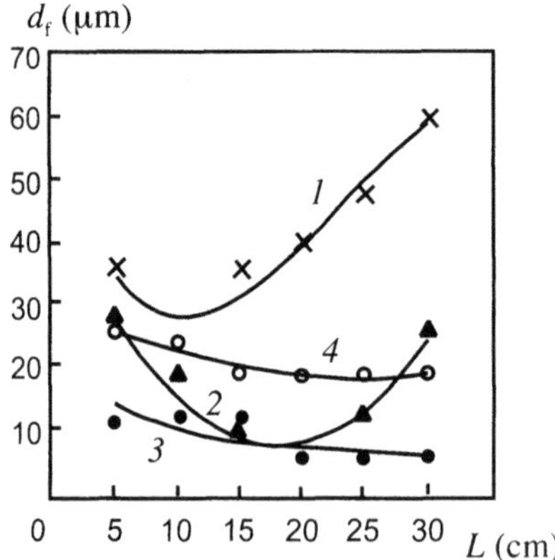

Fig. 4.9. Fiber diameter dependence on the distance between spray head and substrate. Designations are the same as in Fig. 4.7

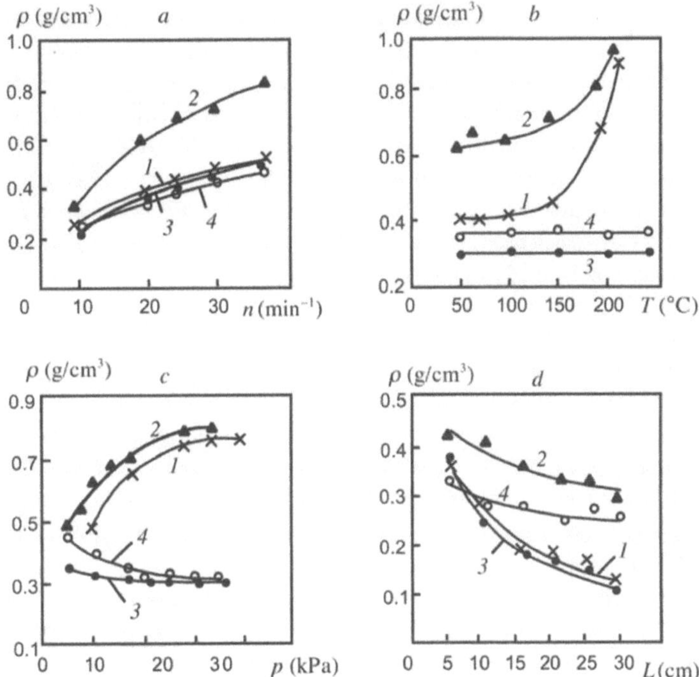

Fig. 4.10. Density of fibrous materials versus (**a**) worm rotational speed, (**b**) spray air temperature, (**c**) pressure, and (**d**) distance between the extruder head and the substrate. *1*, LDPE; *2*, LDPE + SF; *3*, PA; *4*, PA + SF

unfilled polymers is lower than that of filled ones containing ferrite particles whose density is five to seven times higher than that of the binder. This is especially evident for melt-blown materials based on LDPE (curves *1* and *2*). The negligible difference in the density of materials based on PA (*3* and *4*) is explained by only an 8% difference in density between filled and unfilled PA, whereas for LDPE this difference is more than 18%.

An increase of the spray gas temperature (Fig. 4.10b) affects the density of materials based on LDPE (*1* and *2*) at $T > 150°C$, after which further heating leads to increasing stickiness of fibers and the number of welding "bridges" between them. Fiber welding makes their packing denser inside melt-blown materials. PA-based material density does not practically change within the temperature range studied (up to 250°C). This is natural if the melting temperature difference between LDPE (110°C) and PA (250°C) is considered. The density of filled fibrous materials is higher than that of unfilled materials because, first, the filler is heavier than the binder and, second, filled heavy fibers in the viscous-flow state cool down more slowly than unfilled fibers and so pack more densely on the substrate.

The dependencies of fibrous material density on spraying gas pressure (Fig. 4.10c) can be divided into two groups. For LDPE-based materials, all other conditions being equal and averaged by n, T, L, the exponential growth of density is unique. The reason is the formation of welding bridges whose number increases as the pressure increase in the gas flow reaches its limit, which corresponds to the maximum productivity of the spray head and fiber temperature when the tackiness is the highest. PA-based materials show negligible density reduction with pressure increase because of fiber cooling in the gas flow and weakening of bonds in welded bridges. As a result, a looser and less dense structure is formed.

The general trend of density dependence on spray distance (Fig. 4.10d) consists of an exponential reduction in density at increasing distance from the head to the substrate. This happens because of fiber cooling and weakening of cohesion between fibers for greater L values. Limiting density value at $L \to \infty$ corresponds to that of wool made of non-bonded fibers. Presumably, this is the reason that the curves of unfilled PA and LDPE practically coincide despite the difference in the specific mass of these polymers. Filled materials are heavier and their dependencies $\rho = f(L)$ differ significantly.

The regularities cited reflect general trends in the formation of material structure by the melt blowing technique. The scatter of structural parameters governed by polymer composition is found within those trends that determine the technological potentialities of producing fibrous materials for different purposes.

Concluding the review of the technological aspects of melt-blown material formation, it is to be emphasized that the technology provides for multilevel control of material structure from the molecular level to their fibrous texture. Fibers can be modified at any operating stage of the melt blowing process, including dispersion by gas flow and fiber deposition on the substrate. Owing to this fact, the range of technical materials has been enriched by the multitude of multifunctional fibrous composites with unique properties and fields of application.

5. Specific Properties of Melt-Blown PFM

The specific structure of melt-blown materials conditions their unusual service characteristics, which are determined by porosity, the large specific area of PFM, and the high physicochemical activity of the fiber surface layer. Such activity is caused by the specifics of the melt blowing process when fibers are formed from the polymer melt at the boundary of thermal-oxidation destruction. Melt blowing leads to the generation of a spontaneous polarizing charge, which imparts unexpected properties to PFM. Some of them are considered in this chapter.

5.1 Physicochemical Characteristics

The melt blowing process in air is accompanied by thermal-oxidation destruction of the processed polymers. These processes run most intensively in the fiber surface layer due to which is saturated with radicals and active groups (branched end, unsaturated, polar oxygen-containing - hydroxyl, carbonyl, etc.). The processes of thermal-oxidation destruction multiply, and filled polymers acquire specific traits at melt blowing.

As a result, melt-blown PFM do not achieve the characteristics of elevated reactivity for pure polymers. This is manifested in strengthened adhesive interaction between polymer fibers and modifiers (gas, liquid, and solid phase) and in increased adsorptivity of melt-blown materials filtering admixtures from filtered media (see Chap. 9). This creates the conditions for marked improvement of the sorptive capacity of filtering materials. Hereinafter, some -experimental data are cited on the transformation of polymeric physico-chemical structure during melt blowing.

Reduction of macromolecular length is an inevitable outcome of polymer spraying in the viscous-flow state. The melt indexes (MI) of pure thermoplastics and melt-blown materials are given in Fig. 5.1.

MI was determined following Russian State Standard 11645-73 using a capillary viscosimeter (nozzle diameter: 2.005 ± 0.005 mm; load: 11.7 N; time of melt endurance in heated instrument: 5 min). An analysis of the dependencies in Fig. 5.1 proves the supposition that PA fluidity within the T interval 190–250°C increases with temperature rise faster than that of LDPE. As seen from the figure, the MI of LDPE increases two orders as a result of

MI (g/10 min)

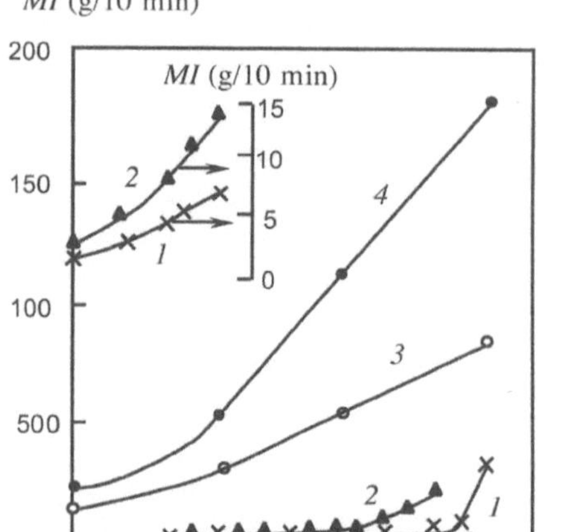

Fig. 5.1. Temperature dependencies of MI of pure thermoplastics (1, 2) and melt-blown materials (3, 4): *1*, LDPE; *2*, PA; *3* and *4*, melt-blown LDPE fibers with diameter 70–80 and 30–40 µm, respectively

melt blowing. The MI of melt-blown fibers is higher, the thinner the fibers. This is evidence of molecular mass reduction of the polymer processed by the melt blowing method, because molecular breakage via oxidation occurs more actively in thin fibers. Curve *1* reflects the LDPE tendency to flow at melt blowing: 320°C is the minimum temperature at which fibers are formed; 340°C is optimum temperature of product manufacturing; 350°C is the temperature of thin-fiber material formation.

Oxidation of a polymer 1 begins upon extrusion and is accompanied by formation of oxygen-containing hydroxyl $-$OH (alcohol) and carbonyl $>$C$=$O (ketone, aldehyde, acidic, ester) groups in macromolecules [1]. Moreover, the authors [2] have identified peaks in the IR spectra of melt-blown LDPE samples corresponding to vinylidene, vinyl, and *trans*-vinylene groups (Fig. 5.2).

A number of fillers promote polymer oxidation during melt blowing; among them are barium and strontium ferrite powders ($BaO \cdot 6Fe_2O_3$ and $SrO \cdot 6Fe_2O_3$) used in producing magnetic filtering PFM (see Chap. 8).

As structural investigations have shown, these finely dispersed magneto-solid ferrites consist of single-domain magnetic particles of 1.8×2.7µm average size (Fig. 5.3a). These ferrite particles that experience magnetic attraction, having developed surface and excess surface energy, are prone, however, to combine in aggregates consisting of two to four particles. The characteristic dimensions of such aggregates are 5.5×8.7µm (Fig. 5.3b).

Aggregates of particles formed during ferrite powder transportation and storage, usually display high strength. They do not collapse during extrusion with a polymer melt, either in granulation or in forming melt-blown PFM.

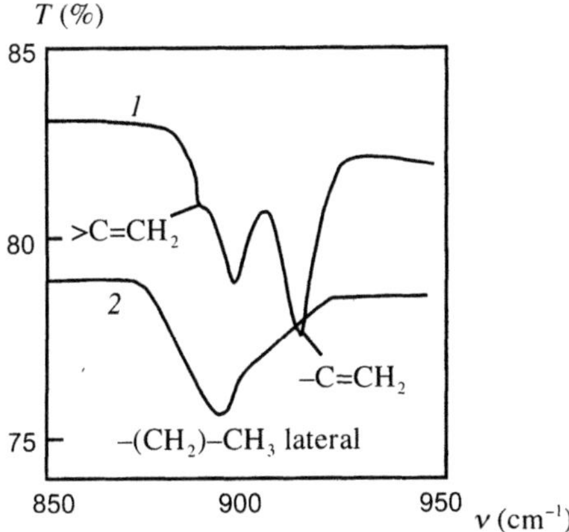

Fig. 5.2. IR adsorption spectra of LDPE films formed from melt-blown fibers (1) and pure polymer (2)

According to X-ray spectral microanalysis (XSMA) data, magnetic particle aggregates are distributed discretely in the bulk of the polymer granules processed into fibers (Fig. 5.4a). Analogous concentration curves of ferrite distribution are observed in fiber cross-sectional investigations of magnetic PFM (Fig. 5.4b). Particle aggregates of about 5 µm (along the peak half width of XSMA concentration curves) are encapsulated inside the binder and do not migrate to the fiber surface.

Chemical interaction between the ferrite and the polymer melt occurs during extrusion of the polymer composition containing dispersed ferrite filler. Ferrite-filler-induced thermal and mechanical oxidation processes in the polymer binder confirm this, in particular, during magnetic composite melt blowing into PFM. Thus, the comparison of the IR spectra of PFM samples shown in Fig. 5.5 has made us conclude that degree of the PE oxidation in ferrite-filled material is much higher than in unfilled analogues produced under identical conditions.

As mentioned before, oxidation of polymers is accompanied by the formation of acidic, aldehyde, ketone, alcohol, ester, and other side and end groups in polymer macromolecules. The relative intensity of the adsorption bands corresponding to valence (ν) and deformation (δ) fluctuations of these groups ($\nu_{C=O}$ at 1730–1750 cm^{-1}, ν_{C-O} at 1080–1130 cm^{-1} and ν_{C-O}, δ_{O-H} at 1250–1380 cm^{-1}) is considerably higher in the IR spectrum of ferrite-filled PFM.

It was shown earlier [3–8] that in the oxidation of PE in contact with active metal substrates (steel, copper, zinc, etc.) forming ion oxides, the oxidation velocity in the metal-bordering polymer layers is much higher than in those distant from the substrate. An accelerated accumulation of oxygen-containing groups in the PE layers adjoining the substrate is accompanied

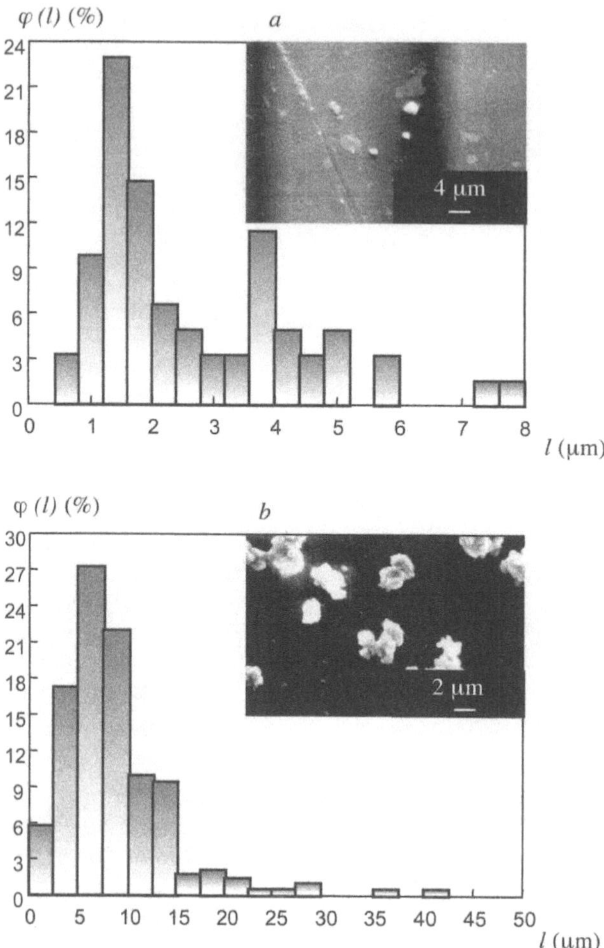

Fig. 5.3. Electron microphotographs and histograms showing strontium ferrite particle distribution lengthwise: (**a**) separate particles dispersed by a supersonic effect; (**b**) aggregates

by the formation of metal carboxylates as a result of the interaction between the oxidized polymer melt and the metal surface. Thus "dissolved" metal in the ionic state migrates into the polymer to a distance of tens of micrometer, and the oxygen liberated from metal oxides might participate in oxidation reactions.

Some metal fillers (copper powder, etc.) also intensify oxidation reactions in the polymer binder, especially around filler particles in the initial stage of thermal treatment [5].

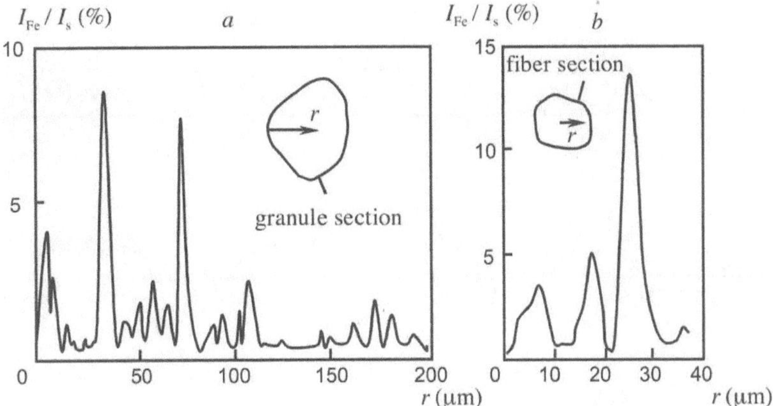

Fig. 5.4. Concentration curves of strontium ferrite distribution in X-ray characteristic beams of FeK_a along the cross-sectional radius of a composite material (PE + 15 mass% of ferrite): (**a**) granule; (**b**) fiber. I_{Fe} and I_s - radiation intensities of the sample and standard, respectively

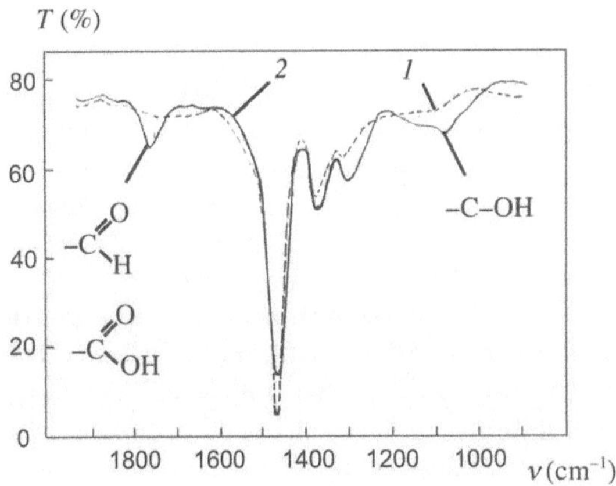

Fig. 5.5. IR adsorption spectra of melt-blown materials: *1*, LDPE; *2*, LDPE filled with strontium ferrite

Analogous processes obviously occur when complex iron oxides (including ferrites) come in contact with polymer melts. Some model experiments were carried out to elucidate their mechanisms.

LDPE film was oxidized at 110–140°C for 5 days on a substrate of a strontium ferrite permanent magnet. During reference testing, LDPE film was endured under conditions similar to a Teflon substrate.

The chemical structure and composition of the films were investigated using XPS, AFM, XSMA, DSC, and IRS.

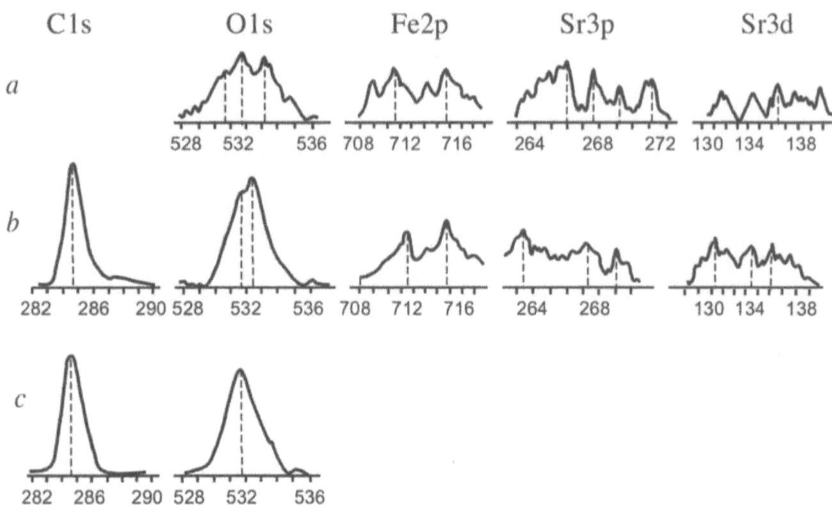

Fig. 5.6. X-ray electron spectra of sample surfaces: a, ferrite substrate (strontium ferrite); (**b**) and (**c**), oxidized LDPE films (t = 5 days, T = 110–140°C); (**b**) contacted ferrite substrate; (**c**) reference sample

X-ray electron spectra of strontium ferrite (mixed oxide $SrO\cdot6Fe_2O_3$) display broadened split spectral lines of Ols, Sr3p, Fe2p (Fig. 5.6a). The constituent of the Ols line with a maximum at bond energy of 533.1 eV probably belongs to oxygen-containing admixtures, and peaks at 531.8 and 530.6 eV belong to oxygen in metal compounds SrO and Fe_2O_3 [9, 10]. In the broadened Fe2p line, one can see maxima at 710.6 eV (characteristic of Fe_2O_3 in mixed oxides) and 714.8 eV. Spectral lines Sr3d and Sr3p are multiple: the main Sr3d peak at 135.5 eV and peaks at 266.0, 267.6, and 271.1 eV. Tabular values of bond energy for SrO are 135.1 (Sr3d) and 269.2 eV (Sr3p), respectively [9]. Broadening and multiple splitting of the lines in X-ray electron spectra of strontium ferrite are obviously connected with the magnetic properties of the compound.

Cls and Ols lines are present in the X-ray electron spectrum of the test film surface oxidized in contact with a ferrite magnet (as in the Reference, Fig. 5.6b) along with weak Sr3p, Sr3d, and Fe2p strips (Fig. 5.6c). The Ols line has a relatively strong intensity and includes a new high-energy component at 532.2–532.5 eV. Its location on the energy scale coincides with the Ols line in metal carboxylates (532.0–532.5 eV), particularly, of $Sr(OOCCH_3)_2$ [9]. This Ols component is absent in ferrite spectra and those of the reference sample. In the latter case, the main peak of Ols is found at 531.8 eV, which is typical of high-molecular weight compounds containing carbonyl and hydroxyl groups [9].

Of interest is the Cls line broadening (within the test sample in contrast to the reference) and the appearance of a low-intensity shoulder added to the main peak of carbon at 288-290 eV. Probably, this can be explained by the location of a maximum in the Cls line of metal carboxylate spectra at 289.0-289.3 eV [9].

So, the presence of Sr3p, Sr3d, and Fe2p lines in the spectra of experimental samples in the regions typical of bivalent strontium compounds and trivalent iron is proof of ionic transfer of these metals (or their compounds) into the polymer film.

The Fe2p band is broadened and has maxima at 714.8 and 711.6 eV. According to the literature [9], the Fe2p line with a maximum at about 711.4–711.5 eV conventionally belongs to chemically pure Fe_2O_3. There are, however, data evidencing that the Fe2p line in the spectra of carboxylate iron (III) compounds is located in the 714 eV region (e.g., the Fe2p line of iron oxalate corresponds to 713.8 eV).

Multiple bands Sr3d and Sr3p contain components typical for SrO (Sr3p: 269.2 eV; Sr3d: 135.1 eV) and complex strontium oxides (Sr3d: 133.7 eV) [9]. However, there are also low-energy components (Sr3d: 130.2 eV; Sr3p: 263.3 eV) which are absent in initial ferrite spectra (Fig. 5.6a). They evidently, belong to some other, possibly metal-polymer compounds of strontium.

As follows from the previous analysis of X-ray electron spectra of model samples, the dissolution of the strontium ferrite surface layer takes place on contact with oxidized molten PE. Sr and Fe oxides are solvated by oxidized macromolecules and by products of their thermal destruction and transfer into the ferrite-bordering polymer layer. The oxides start to crystallize in the polymer matrix during melt cooling. Along with this, part of the oxides enter into a chemical reaction with low- and high-molecular weight components of oxidized melt- forming metal-containing carboxylate compounds.

This conclusion is supported by the results of the system under study that were deduced by other physicochemical methods. Thus, AFM images of the test sample (oxidized in contact with ferrite) and the reference display apparent differences (Fig. 5.7). The surface layer material of the reference film is mostly homogeneous in its rigidity. Darker and lighter portions located, as a rule, within spherulite boundaries (Fig. 5.7c) are, probably, the result of negligible changes in the degree of polymer crystallinity across the sample surface area. Characteristic spherulite dimensions are 1–1.5 μm.

The surface topography of the test sample (Fig. 5.7a) has another relief and is mostly composed of submicron structural formations. The melt crystallization process turned out to be faster here, presumably because of altered heat and physical properties of the material and the formation of additional crystallization centers. The phase contrast image (Fig. 5.7b) clearly shows more rigid phase inclusions of a different nature in the polymer matrix (light portions). These inclusions are 0.1–0.5 μm in size and can be related, most likely, to Fe and Sr oxides. The oxides are transferred from the ferrite surface via "dissolution" by the polymer melt and crop up during crystallization

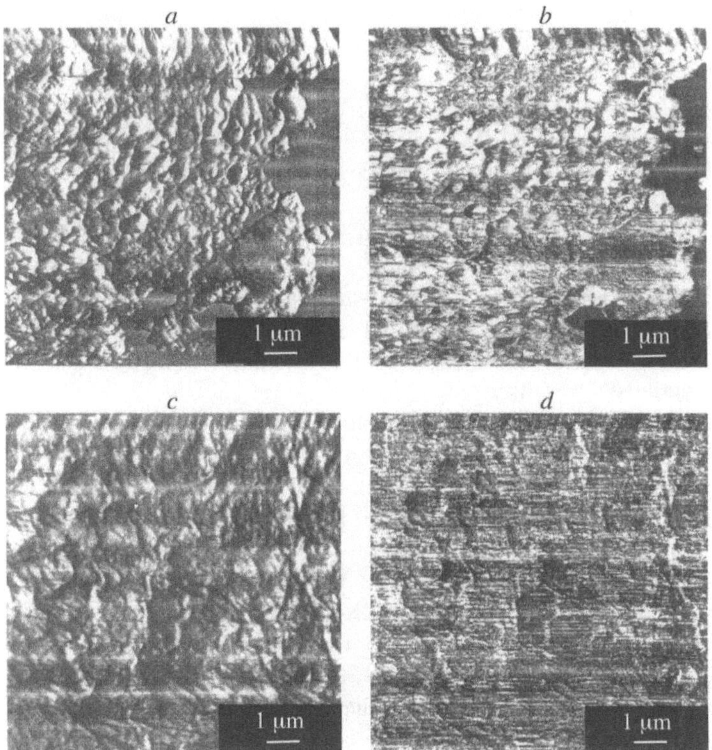

Fig. 5.7. AFM images of surfaces oxidized in melt LDPE film in topographic exposure regimes ((**a**), (**c**)) and phase contrast in the rigidity parameter ((**b**), (**d**)): (**a**), (**b**), sample contacted by a ferrite magnet; (**c**), (**d**), reference sample

in the polymer surface layer as a colloidal phase. Evidently, just these new phase centers influence (as described before) the behavior of polymer structural formation during crystallization.

The fact of varying thermophysical characteristics, in particular, the change of PE phase transition temperatures upon contact oxidation on a ferrite substrate or combined processing with a highly dispersed ferrite filler, is supported by DSC data (Table 5.1). The melting peak of these materials on DSC curves is expanded and is located in the region of higher temperatures (113–116°C). Positive (compared to pure LDPE samples) melting ΔT_{m} and crystallization ΔT_{cr} temperature increments indicate the formation of a new metal-polymer phase in the materials under study.

Ferrite diffusion into the polymer has been confirmed by X-ray spectral microanalysis (microprobe Comeka MS-46, FeK$_\alpha$ and SrK$_\alpha$ radiation) by X-ray scanning of the film cross-section. According to XSMA data (Fig. 5.8), the penetration depth of "dissolved" metals into the polymer layer in model experiments is 5–7 μm for Fe and 8–10 μm for Sr. There are two maxima on

Table 5.1. Thermophysical characteristics of LDPE-based samples, according to DSC data

Sample type	T_{m} (°C)	T_{cr} (°C)
Filler-free granulate	110	94
Filler-free melt-blown material	109	95
Film oxidized on ferrite substrate[a]	113	98
Granulate with ferrite filler[a]	116	102
Melt-blown material with ferrite filler[a]	114	98

[a]Strontium ferrite $SrO \cdot 6Fe_2O_3$

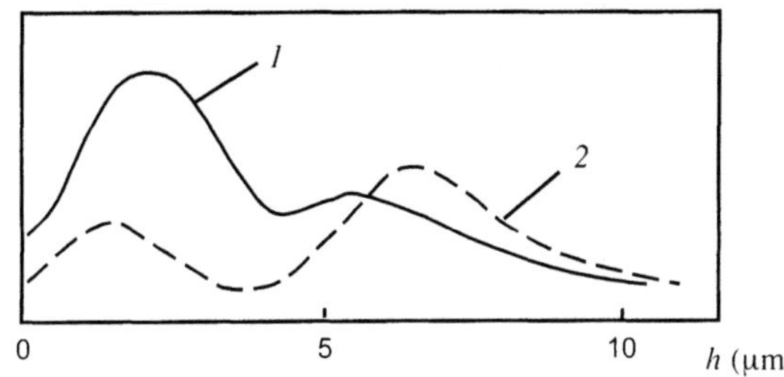

Fig. 5.8. Concentration curves of Fe (1) and Sr (2) depth distribution across the surface layer of melt-oxidized ($t = 5$ days) LDPE film in contact with a ferrite magnet (according to XSMA)

the concentration distribution curves of these elements across the depth of the film surface layer. The first of them evidently corresponds to cropped-up oxides; the second, to more deeply penetrated organometallic compounds. Note that strontium oxide incorporated into the ferrite composition interacts with the polymer melt more actively than ferric oxide. Due to this, dissolved SrO and its products of interaction with the melt are accumulated faster and penetrate deeper into the polymer layer.

Finally, interesting information on the strontium ferrite interaction with PE melt was furnished by the IRS method (Figs. 5.9, 5.10, Table 5.2). Indeed, the processes of PE thermal and oxidation destruction accompanied by the accumulation of unsaturated and oxygen-containing groups in macromolecules proceed faster and deeper in the polymer-ferrite pair than in the reference polymer-neutral substrate (Teflon). In the first case, the relative intensity of bands related to the groups indicated in the IR spectra are considerably higher in both the polymer film bulk (Fig. 5.9) and in the surface layers (Fig. 5.10).

In addition, a number of new rather intensive bands appear in the IR internal reflection spectra of the test sample, predominantly from the ferrite-contact side in contrast to the reference (Fig. 5.10, Table 5.2). The first group

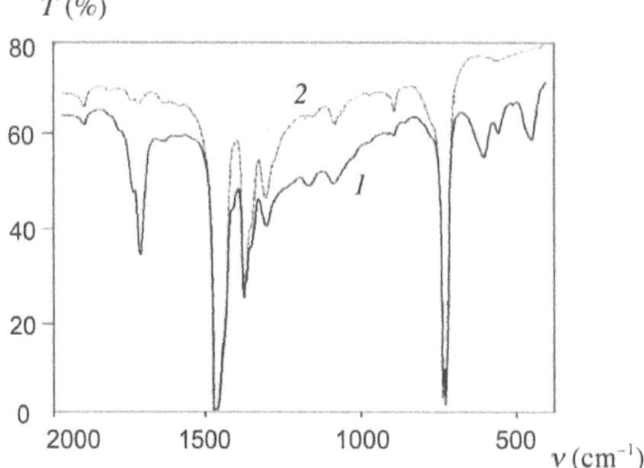

Fig. 5.9. IR absorption spectra of melt-oxidized LDPE films ($t = 5$ days): 1, sample contacted by a ferrite magnet; 2, reference sample

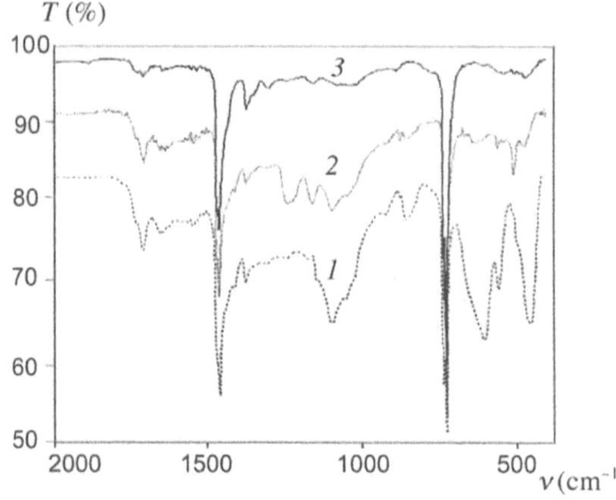

Fig. 5.10. IR internal reflection spectra of melt-oxidized LDPE films ($t = 5$ days, $T = 110$–$140°C$): 1 and 2, test sample from ferrite contact side (1) and back side (2); 3, reference sample

of bands at 1658, 1550, and 1397 cm^{-1} is unique to valent (asymmetrical and symmetrical) fluctuations of the C=O bond within the carbonic acid layers [6–8, 10–12]. The second group at 448, 550, and 598 cm^{-1} is specific for valent fluctuations of Me=O and Me-O bonds in Fe_2O_3 and SrO oxides [12–14].

Thus, IRS data confirm the hypothesis formulated on the mechanism of ferrite and LDPE interaction through dissolving Sr and Fe ion oxides by the oxidized melt and their interaction with the latter leading to the formation of carboxylate type compounds. As a result, a new metal-polymer phase is formed that has higher T_m and T_{cr} than pure LDPE, which plays an independent part in electropolarizing processes. Particularly, this phase contains

Table 5.2. Characteristic bands of valent (ν) and deformation (δ) fluctuations of unsaturated and oxygen-containing groups in IR internal reflection spectra of oxidized LDPE films ($t = 5$ days, $T =110\text{--}140^\circ$C)

Observed frequencies (cm^{-1})[a]			Band classification according to [1, 11–14]
Test sample		Reference sample	
Ferrite-contacted side	Back side		
$3300_{\text{av wide}}$	$3300_{\text{av wide}}$	–	$\nu(\text{O}-\text{H}_{\text{assoc.}})$
1745_{av}	1745_{av}	1745_{w}	$\nu(\text{C}=\text{O})$
1715_{av}	1714_{av}	1714_{w}	$v(\text{C}=\text{O})$
1658_{av}	1658_{w}	$1656_{\text{v.w}}$	$\nu(\text{C}=\text{C})$, $\nu_{\text{as}}(\text{COO}^-)$
1550_{av}	1550_{w}	–	$\nu_{\text{as}}(\text{COO}^-)$
1397_{av}	1397_{av}	–	$\nu_{\text{s}}(\text{COO}^-)$
1240_{av}	1239_{av}	–	$\nu(\text{C}-\text{O})$
1170_{av}	1160_{av}	1160_{w}	$\nu(\text{C}-\text{O})$
1093_{st}	1094_{av}	–	$\nu(\text{C}-\text{O})$
1050_{av}	1050_{av}	$1060_{\text{w.wide}}$	$\nu(\text{C}-\text{O})$
920_{av}	$925_{\text{w.shoulder}}$	–	$\delta(=\text{CH}_2)$
–	876_{w}	890_{w}	$\delta(=\text{CH}_2)$
853_{av}	854_{w}	–	$\delta(=\text{CH}-)$
598_{st}	–	–	$\nu(\text{Sr}=\text{O})$
550_{av}	–	–	$\nu(\text{Fe}=\text{O}, \text{Fe}-\text{O})$
448_{av}	–	–	$\nu(\text{Fe}=\text{O}, \text{Fe}-\text{O})$
$497_{\text{w.shoulder}}$	503_{av}	–	$\delta(=\text{CH}-)$

[a]st: strong; av, average; w, weak; assoc, associative; s, symmetrical; as, asymmetrical.

additional spatial localization centers of charge carriers, which implement strong trapping (see Sect. 5.2).

Furthermore, a metal-polymer phase formed at the ferrite filler-binder interface affects the topography of the magnetic field generated by the filtering material and makes a certain contribution to the mechanism of filtering processes conditioned, in particular, by the phenomena of magnetic and heterocoagulation of contaminants (see Chap. 8).

5.2 Electret Charge in Melt-Blown Materials

Unusual physicochemical characteristics of PFM melt-blown materials arise from chemical modification of the surface layer and also from physical modification of the fiber permolecular structure. Fiber formation is accompanied by severe thermal and mechanical effects on the polymer melt in a viscous-flow state, which brings about a spontaneous electric charging of the fibers transforming them into the *electret state*. An electret is a dielectric that preserves its electrified state for a long time and creates an electrical field in the environment like permanent magnets create magnetic fields.

In [2], the parameters of electrical polarization have been estimated in melt-blown materials obtained by LDPE deposition on a grounded metal substrate under various melt-blowing regimes. The speed of extruder worm rotation n, the temperature T of spray head, and the pressure p of air supplied to the head were varied. Before measurements, the samples were calibrated and conditioned according to Russian State Standard (RSS) 6433.1-77. After that, the efficient surface charge density σ_{ef} (ESCD) of the electret was estimated by the compensation method, according to RSS 25209-82. Then, the activation energy W, relaxation time of residual charge τ, trapping section S, and the frequency of liberation attempts ν_o of the carriers from the trapping center were determined by the method of thermally stimulated depolarization (TSD).

TSD current spectra of all samples investigated show a characteristic peak at $102\pm5°C$ corresponding to negative charge relaxation (Fig. 5.11). After 160 days in a humid atmosphere, the samples showed a reduction in peak intensity of 40%. The presence of a single peak provides the basis for a supposition that a volume charge is formed in the sample by the carriers captured by localized traps of a similar localizing depth in a forbidden band [15].

The main parameters describing the capture and recombination of charge carriers are given in Table 5.3. Analysis leads to the following conclusions.

A bipolar polarizing charge is formed in the samples when $\sigma_{ef} = 0.39$–0.43 nC/cm^2 and $W = 0.83$–0.85 eV. It is sufficiently stable, which is supported by negligible changes in charge characteristics upon sample storage. An increase in σ_{ef} is probably due to relaxation of the negative sign. That is

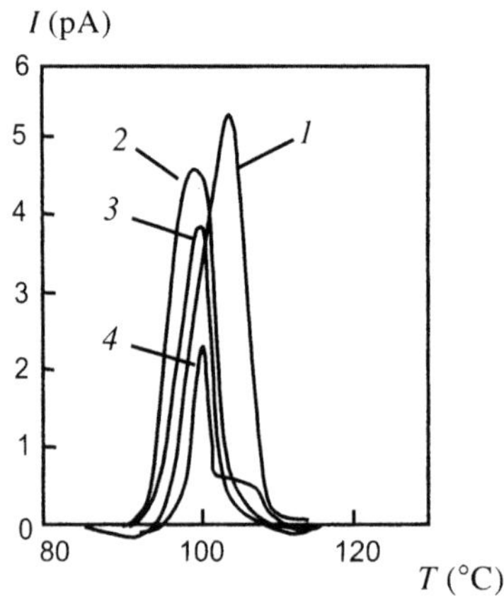

Fig. 5.11. TSD current spectra of LDPE-based fibrous material taken during 1 day (1), 30 (2), 60 (3) and 160 (4) days after sample fabrication

Table 5.3. Variation of charge state characteristics of melt-blown materials with time

Time from sample fabrication till testing (days)	ESCD σ_{ef} (nC/cm^2)	Activation energy W (eV)	Relaxation time $\tau(10^5$ s)	Capture section $S(10^{-20}\mathrm{m}^2)$	Frequency of liberation attempts $\nu_o(10^{10}$ s$^{-1})$
1	0.39	0.84	2.7	1.9	8.0
30	0.41	0.83	2.5	3.7	1.5
60	0.43	0.83	3.1	5.7	2.3
160	0.42	0.81	3.4	4.5	1.8

why the relaxation time of the total charge increases, and the frequency of escape attempts is reduced with storage time under constant conditions.

The electret charge is identified by three interconnected criteria [16]: (1) the presence of bulk electric charge in the sample; (2) its prolonged preservation, and (3) generation of TSD currents in the "electrode-sample-electrode" circuit upon electrode shorting and sample heating. The results cited prove the presence of these criteria in the samples under study and also point to the fact that the melt blowing technique leads to the generation of an electret state in nonpolar and weakly polarizing LDPE without using any electrical sources [2].

The technological regimes of the melt blowing process influence the electret charge value in fibrous materials. In Fig. 5.12, TSD current spectra of samples formed under different conditions are presented, where n is the worm rotational speed, p is the pressure of spray air, and T_1 is the melt temperature in the extrusion head. It is evident that as the influence of each of these factors intensifies, the current value also increases.

It was shown earlier (see Fig. 4.8) that increases in T and p lead to reduced fiber diameter d_f. Simultaneously, σ_{ef} and W values increase (Table 5.4). The growth of n (at T, p = const) is also accompanied by σ_{ef} increase in spite of d_f increase (see Fig. 4.7), although an increase in W is not noticed. So, it can be anticipated that with an increase in n, the degree of macromolecular destruction also increases, and consequently, new trapping centers appear. Presumably, an increasing number of trapping centers affects σ_{ef} more strongly than a d_f increase.

The TSD current spectra presented in Fig. 5.12 are a result of the superposition of currents generated from the thermal relaxation of fiber volume and surface charges. These charges possess different signs, and their value is correlated with the material structural parameters and, consequently, with the technological regimes of the melt blowing process.

The electrification mechanism of melt-blown materials can be presented in the light of the previously cited experimental data.

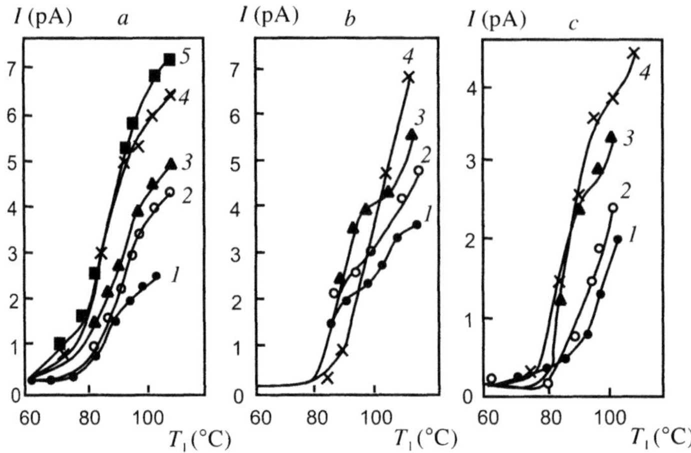

Fig. 5.12. TSD current spectra of fibrous materials formed under various technological regimes: (**a**), $n = 20$ min^{-1}, $p = 30$ kPa, $T_1 = 310, 340, 360, 380$, and $390°$C for curves 1–5, respectively; (**b**), $T_1 = 390°$C, $p = 30$ kPa, $n = 4, 20, 25$, and 30 min^{-1} for curves 1–4, respectively; (**c**), $T_1 = 390°$C, $n = 20$ min^{-1}, $p = 10, 20, 30$, and 40 kPa for curves 1–4, respectively

Table 5.4. Characteristics of the charge state of melt-blown materials depending on their technological regimes

Forming regimes		Charge state characteristics				
Extrusion parameter	Parameter values	σ_{ef} (nC/cm^2)	W(eV)	$\tau(10^5$ s)	S (10^{-20} m^2)	ν_{o} (10^{10} s^{-1})
	310	0.13	0.82	0.1	19.0	7.9
	340	0.16	0.81	4.2	3.8	1.5
$T_1(°$C)	360	0.25	0.84	3.1	5.1	9.0
	380	0.27	0.83	4.7	5.3	7.1
	390	0.55	0.85	5.0	6.0	8.2
	14	0.3	0.87	0.9	3.0	0.5
n(min^{-1})	20	0.35	0.82	1.6	3.7	3.2
	25	0.37	0.86	3.1	4.0	2.1
	30	0.43	0.88	4.2	6.9	1.8
	10	0.14	0.83	1.9	0.2	0.5
p(kPa)	20	0.22	0.82	3.3	3.7	1.5
	30	0.42	0.84	4.2	2.8	2.8
	40	0.55	0.85	7.1	4.8	1.6

Trapping centers in solid polymers are subdivided by the criterion of spatial localization of charge carriers into coulombic centers of attraction, repulsion, and neutral ones [15]. Capture cross sections of coulombic centers vary from $S = 10^{-5}$ m^2 (attraction) to 10^{-25} m^2 (repulsion), and S of neutral centers is found between those values. Strong capture trapping centers are characterized by the frequency of carrier escape attempts $\nu_o = 10^8$–10^9 s^{-1}; those with weak capture are 10^{13}–10^{14} s^{-1}. From the analysis of the data presented in Tables 5.3 and 5.4, it is evident that charge carriers in melt-blown materials localize in neutral centers, which realize strong capture.

Elevated temperatures at which polymers are sprayed approach their destruction thresholds and initiate breakage of the main chain bonds in macromolecules together with the formation of radicals. The breakage is preceded by thermal-fluctuation-induced elongation of bonds. At sufficiently large elongation, self-ionization of macromolecules can occur, when bound electrons "soak" through the potential barrier of the macromolecular force field thus realizing the tunnel effect [17]. When a polymer undergoes melt blowing, the probability of self-ionization is highest during intensive stretching and dispersion of overheated fibers in a compressed air flow. The number of ionized macromolecules increases with the rising T_1, which is evident from the increase in σ_{ef}, S, and τ (Table 5.4). The ionized state of macromolecular patches is stable enough because correlation between T_1 and frequency ν_o is not observed.

A lot of macromolecules in a viscous-flow polymer mass have "innate" defects. These are oriented in a certain manner atomic groups with either constant or induced dipolar moments. Such defects can act as structural traps [18]. When the polymer mass in a viscous-flow state contacts the metal parts of the spray head, charge carriers injected from the metal localize on these centers. This is the reason that the increase in contact interaction intensity is immediately recorded as a TSD current gain in melt-blown material (Fig. 5.12). When n increases from 20 to 30 rpm (Table 5.4), the frequency ν_o of charge carrier escape attempts is reduced. This agrees with the proposed hypothesis for the polarization mechanism of fibrous materials and supports the high probability of macromolecular self-ionization and mechanical destruction during melt blowing.

Charges acquired by fibers at the high-temperature processing stage are rather stable. They do not manage to escape from the traps during spraying due to fast cooling of the fibers and remain "frozen" in their localized states. It is seen from Table 5.4 that all of the parameters of the sample charge state increase with a p rise. Probably, the processes of macromolecular ionization and mechanical destruction go on during melt spraying. It is natural that the processes of polymer-metal contact breakage make an essential contribution to the mechanism of fiber polarization. This takes place when the gas flow tears fibers forced through the die away from the head. With a rising p, dispersed fiber fragments become shorter, the frequency of their separation increases, and the polarizing charge on the fibrous material grows.

The hypothesis for carrier localization on readily polarizing macromolecular groups with double bonds has one more confirmation. Carrier capture just by neutral centers conditions high stability of the electret charge of melt-blown materials with time.

The processes of carrier entrapment and polarizing charge relaxation continue during fibrous mass deposition on the substrate. Their intensity is, however, two orders lower than at the high-temperature stage of the process. Charge transfer localizes on fiber contact sites. Secondary polarization processes induced by electrical fields of neighboring fibers proceed till polymer hardening.

The filler effect on polarizing charge generation in melt-blown materials can be explained by the example of magnetic PFM based on ferrite-filled LDPE.

For pure LDPE samples, only one discharge current peak is typical at 104–105°C. Two additional peaks at 75–77 and 115–120°C are seen on the spectra of samples filled with ferrite (Fig. 5.13). This suggests that three trapping center types appear in the material responsible for spatial localization of charge carriers (Table 5.5). The peak of pure LDPE corresponds to relaxation of the natural polarizing charge of melt-blown material. The low-temperature peak (75–77°C) arising from fiber filling with ferrite is brought about by Maxwell-Wagner polarization. This means that charge carrier accumulation at the filler-binder interface results from the difference in ferrite and LDPE conductivities. The high-temperature peak (115–120°C) is the result of polarizing charge relaxation occurring in the phase of metal-polymer compound, which is formed, according to the hypothesis proposed in Sect. 5.1,

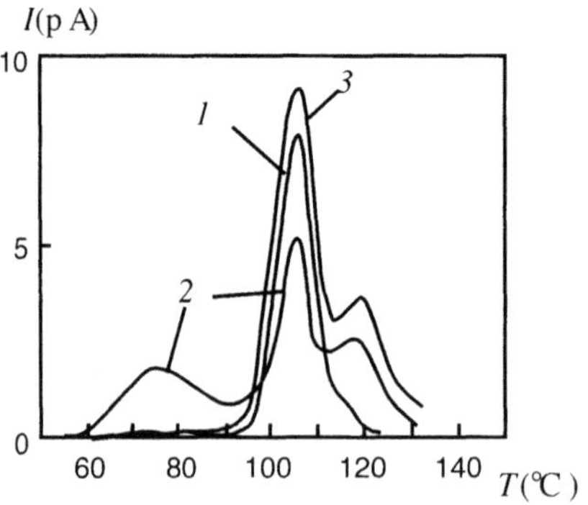

Fig. **5.13.** Temperature dependence of TSD currents in melt-blown materials based on LDPE: *1*, pure; *2* and *3*, filled with strontium ferrite (15 mass %) nonmagnetized and magnetized, respectively

Table 5.5. Characteristics of the melt-blown PFM charge state of LDPE + strontium ferrite (15 mass%) composition

Sample	σ_{ef} (nC/cm^2)	T_{max} (°C)	W (eV)	τ (10^5 s)	S (10^{-20} m^2)	ν_0 (10^{10} s^{-1})
Nonmagnetized	0.41	75–77	0.79	0.3	6.5	21.0
		104–105	0.83	3.2	4.0	5.0
		115–120	0.86	30	0.47	0.02
Magnetized,	0.51	104–105	0.84	1.0	4.0	1.0
$B_{\mathrm{r}} = 0.4$ mT		115–120	0.9	20	0.013	0.032

through a carboxylate mechanism. The presence of a high-temperature peak is a confirmation of this fact [19].

The low-temperature peak corresponds to the temperature of the λ_1 relaxation transition related to the mobility of physical nodes in the molecular net of the LDPE amorphous phase [20]. When the sample is magnetized the peak lessens or disappears completely, and the intensity of the main peak increases. This means that the λ_1 transition in a strong magnetic field is accompanied by migration of charge carriers from low- to high-temperature trapping centers. When material is demagnetized during storage, a low-temperature peak appears again on TSD curves.

Proceeding from this, magnetic melt-blown PFM [21] serve as carriers of both magnetic and electrical fields, which expands their field of application.

6. Fibrous Materials in Filtration Systems

One of the most positive trends in melt blowing technology is the production of fibrous materials for filtering systems. Filtration is the motion of liquids or gases through a porous medium. Liquid or gaseous media are separated from contaminants during filtration. In the process of filtration suspensions or aerosols are separated by porous screens letting liquid or gas pass but keeping solid particles back. Filtration is performed by filters whose major part is the filtering element (FE), which is a porous screen made of a filtering material (FM). A diversity of filtration conditions exist (various sizes and types of contaminants, volumes of media being filtered and velocity of filtration, degree of cleaning, and so on) which determine the use of a wide range of FM, including paper, metal meshes, synthetic and natural fabric and fibers, porous plastics, and powder materials. Melt-blown polymer materials occupy their own merited place within the combinations of synthetic FM and are efficiently used in the techniques of purifying technological and working media, as well as industrial wastes.

6.1 Efficiency of Filtration Systems

The main service characteristics of filters and FEs as well as terms denoting filtering efficiency have been standardized in Russia at the state level [1–4] and are also reflected in international standards [5].

The filtering factor or *β coefficient* corresponds to each contamination fraction with a maximum particle size x:

$$\beta_x = N_1/N_2 \ ,$$

where x is the index of chosen particle size in micrometers, N_1 is the number of particles larger than x per milliliter of liquid before the FE, and N_2 is the number of the same fraction of particles per milliliter of liquid after passing the FE.

In practice, particle size is chosen depending on filter designation and filtering conditions. It is usually of the series 5, 10, 20, 30, and 40 μm, and the β factor is determined for particles of a fixed size. For example, β_{10} is the filtering factor for a particle size above 10 μm.

The filtration ratio is the relation of the amount of particles captured by the FE with sizes exceeding the preset size to the amount of particles with the same size per unit volume of the liquid before the FE:

$$\eta_x = \frac{N_1 - N_2}{N_1} \ .$$

The filtration ratio and the filtering factor β_x are interrelated through the relationship

$$\eta_x = 1 - \frac{1}{\beta_x} \ .$$

The term *efficiency rating* denoted as E_x is often used in scientific and technical literature [5]:

$$E_x = \frac{N_1 - N_2}{N_1} \cdot 100\% = \left(1 - \frac{1}{\beta_x}\right) \cdot 100\%$$

Filtration fineness (sifting) characterizes the capability of a FE to capture contaminant particles of a certain size. The absolute and nominal fineness of filtration are distinguished. *The absolute* filtration fineness δ_{abs} is the maximum particle size (μm) of a spherical artificial contaminant found in filtered liquid. The parameter of absolute filtration fineness is used in practice if the filtering factor of the particle size x is $\beta_x \geq 75$ or the filtration ratio is $\eta_x \geq 0.987$. *The nominal* filtration fineness δ_{nom} is the minimum size (μm) of contaminant particles captured by a FE whose filtering factor is $\beta_x \geq 20$ or filtration ratio is $\eta_x \geq 0.95$.

The FE efficiency by filtration mass indexes is reflected by sifting and transmission coefficients. *The sifting coefficient φ is* the relation of the contaminant mass captured by a FE to that found in an unfiltered working fluid:

$$\varphi = \frac{M_1 - M_2}{M_1} \ ,$$

where M_1 is the contaminant mass in the initial working fluid before a FE; M_2 is the contaminant mass in filtered liquid after the FE.

The transmission (skipping) factor is the ratio of contaminant mass in the filtered working fluid (after a FE) versus the contaminant mass in an unfiltered working fluid (before a FE).

An important parameter that reflects filter efficiency is the pressure differential in a FE, i.e. the pressure difference (Δp) between the FE inlet and outlet when a medium is being filtered. Nominal, maximal, and breaking pressure differentials are discriminated. *The nominal* pressure differential Δp_{nom} is measured on an uncontaminated FE under nominal liquid discharge and a given viscosity value. *The maximal* pressure differential Δp_{max} is the limiting differential value under which FE operation is admissible. *The breaking* pressure differential Δp_{br} brings about either residual deformation or tightness violation of a FE.

The mud capacity of a FE is the artificial contaminant mass of a pre-set particle size captured by a FE from the beginning of filtration till the maximum pressure differential is attained under nominal liquid consumption and a given viscosity. To determine the mud capacity, a system of continuous circulation of contaminated liquid flow through the filter being tested is used at a preset temperature and flow rate. Portions of an artificial contaminant are introduced into the liquid flow before the filter until a limiting pressure differential is reached. The latter is usually indicated either in standards or technical specifications for a certain filter type. Mud capacity is calculated using the formula

$$G = Mn \; ,$$

where M is the mass of contaminant specimen in the filtered media in milligrams, and n is the number of specimens needed to increase the pressure differential from nominal to the limiting pressure.

6.2 Filtration Mechanisms

The filtration of gases or liquids is accompanied by precipitation of suspended particles on a FE. In this connection, general regularities typical of all kinds of filtration and the main features of the purification process of gases and liquids are discussed following.

6.2.1 Mechanisms of Particle Precipitation

Among the major reasons ensuring particle capture by a filtering screen, the following phenomena are named by the classical theory of filtration: direct interception, inertial impaction, diffusion, screening effect, as well as gravitational, electrostatic, and magnetic precipitation [5, 6].

Coarse particles ($> 20\,\mu m$) are entrapped mainly by inertial capture, gravitational precipitation, and screening. *Inertial impaction* consists of the following: a particle bypasses an obstacle (a structural element of FM) in the medium flow, but under inertial forces, it is deflected from the flow direction and collides with the obstacle. The probability of such a capture is higher, the larger the particle size; so inertia has little effect on particles less than $1\,\mu m$ in diameter. *Gravitational precipitation* occurs by the action of the earth's gravity. *Screening effect* is observed when the particle size is too large to pass through the FM pores and channels.

Therefore a filter for eliminating coarse contaminants can be presented as a "mechanical" porous screen that prevents particle penetration but does not impede liquid or gas flow through it. Separation of fine impurities ($< 20\,\mu m$) is much more complicated. The main role here is played by precipitation mechanisms based on impaction and diffusion.

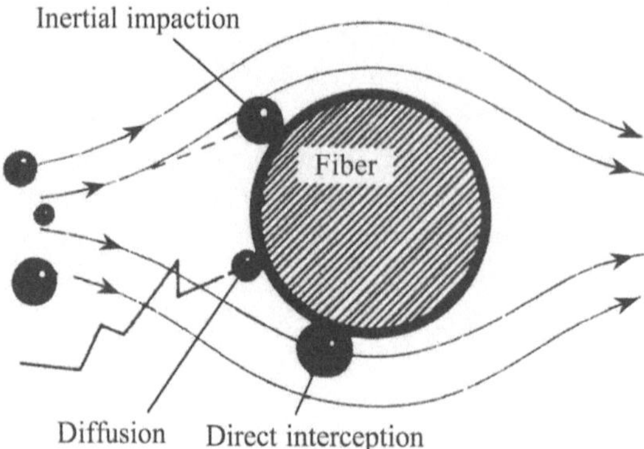

Fig. 6.1. Particle precipitation mechanisms on a single fiber [5]

Direct interception takes place when a particle approaches an obstacle at a distance less than its radius. Particles stick and are retained under the action of van der Waals forces. *Diffusive precipitation* is typical for extremely small particles (below 1 μm), which travel chaotically as a result of Brownian motion and might settle on a FE screen due to random deflection from the flow direction.

The dominance of one or a few precipitation mechanisms is determined by the parameters of the filtered medium, the contaminant particle size, and filtration conditions. A scheme is given in Fig. 6.1 of particle precipitation on a single filament when the main precipitation mechanisms are inertial impaction, direct interception, and diffusion.

6.2.2 Surface and Depth Filtration

There is a distinction between surface and depth (bulk) filtration. *Surface filtration* proceeds mainly by direct interception and screening. Particles with diameter larger than inlet openings of a FM channel are retained on a FM surface. Smaller particles can be blocked in the channels. Adsorptive interactions between contaminants and a FM are usually negligible. The simplest type of "mechanical" filter realizes surface filtration, i.e., a FE serves as a screen that does satisfactorily rough cleaning. During prolonged service of such filters, two effects arise:

1. Gradual narrowing of the FM channel cross section due to partial blocking by particles that leads to "finer" filtration, i.e., the FE starts to capture finer particles because fine particles are retained on channel walls by adsorptive forces (Fig. 6.2).
2. A bed (thick layer) of contaminant particles starts to grow on a FE surface thus forming an additional filtering screen which is blocked by a

mechanism analogous to that previously noted and becomes a finer FE with time (Fig. 6.3).

Another type of mechanical filter presupposes FE employment of considerable thickness, which provides *depth filtration*. In this case, the filtration mechanism becomes more intricate because particle passage through a FE has prolonged and random behavior which furnishes more chances for direct interception and retention of contaminants. Coarser particles are, as a rule, captured by a FE surface, and smaller ones by subsequent layers. The optimal structure of a FE for depth filtration is characterized by a density (porosity) gradient. This is important when impurities have a large size scatter. If the particles are more homogeneous in size, a surface type FE can be used in place of a depth filter. This circumstance sometimes plays an important part in choosing a FE type because the pressure differential on a depth filter is usually larger than that on a surface filter.

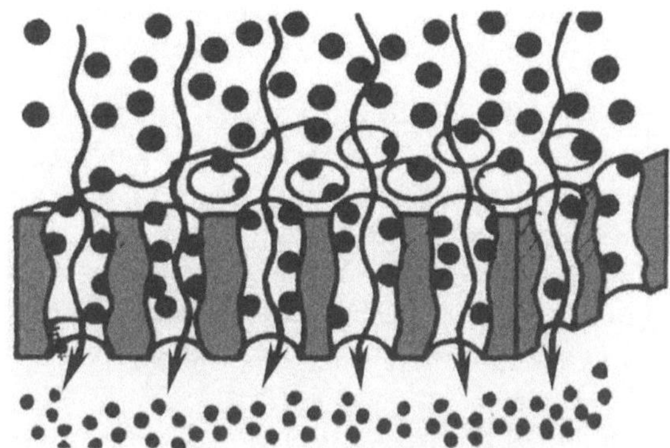

Fig. 6.2. Blocking effect of fine particles retained by a surface FE [5]

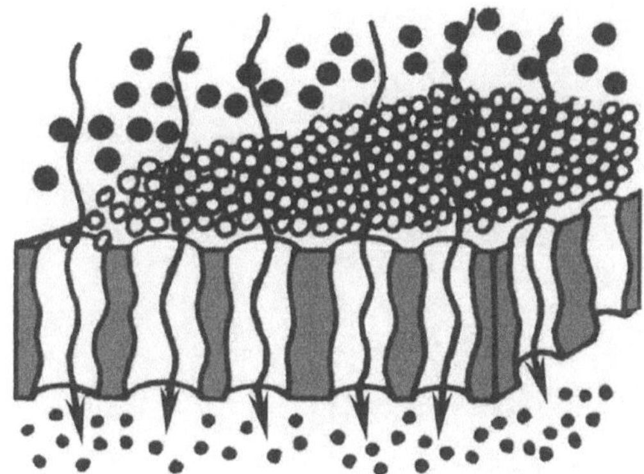

Fig. 6.3. Formation of a bed from particles captured on a FE surface [5]

The total efficiency of a depth type filter is higher than that of a surface filter. In inertial interception of particles, adsorptive forces take part, and, in addition, fine particles in Brownian motion are entrapped by the adsorptive capture mechanism. As a result, a depth filter can intercept finer particles than a surface filter.

An ideal structure of a depth FE presumes a density gradient growing in the flow direction of the filtered medium (Fig. 6.4). Thus, the probability of intercepting fine particles increases during passage through the FE.

There are three types of depth filters: fibrous, porous, and cake-like [5]. A *fibrous FE* consists of a layer of thin fibers oriented randomly relative to each other. Depending on the fiber material and the production technique, their diameter can be from 0.5 µm to a few tens of micrometers. The fibers are mixed and interwoven so that numerous twisting channels are formed in between where particles are entrapped and retained by the mechanisms previously described. In practice, along with synthetic polymers, cellulose, cotton, and glass microfibers are widely used to manufacture fibrous filters. Filtration efficiency depends strongly on fiber diameter. The thinner the fibers, the denser they can be packed, and the shorter will be the path of contaminant particles through a FE.

A fibrous FE layer is usually about 0.25–2.0 mm thick, and the fibers are adhesively bonded. The latter is requisite for a FE structure, including its pores, which ensures, in its turn, stable filtration characteristics. A distinctive feature of melt-blown fibrous materials is welding "bridges" between fibers that impart structural stability and permanence to a FE without using glue compositions. However, there is also an opinion [7] that the filtration efficiency of a FM consisting of nonbonded fibers exceeds that of an FM whose fibers are bonded to each other.

A *porous filter* is like a fibrous FE in providing of capillary penetration of the medium through the pores. The latter are formed either by foaming plastic material (e.g., polyurethane) or by sintering quasi-spherical particles of a solid material (metal, ceramics, polymers [8]).

Fig. 6.4. A model of an ideal structure of a depth fibrous FE

Cake-like (filled) filters are formed in application, and are used mostly to remove solid particles from sizeable volumes of liquids. The simplest cake-like filter is a layer of solid particles, which are placed on a support screen or a mesh. Voids between particles form pores and channels for filtration. Any binders to bond particles are excluded. This structure of the filtering layer makes the filters unsuitable for purifying technical fluids because for this purpose it is important to have a stable, compact FE structure insensitive to vibration. Typical materials for cake-like filters are diatomite, sand, clay, as well as wood and cotton fibers.

6.2.3 Electrostatic Precipitation

Aside from impaction, gravitational precipitation, and inertial and diffusion interception of contaminant particles, electrostatic interaction of unlike charged bodies is the main filtration mechanism for gaseous media. Contaminated gas passage through an electrostatic field is a widely applicable industrial process for removing dust from gas streams. Its main merit consists in realization of significant electrical forces act directly on particles but do not depress the medium stream. This also makes it possible to use some of the advantages of electrostatic precipitation, including the chance of removing particles below submicron size from the gas stream, realization of a high degree of purification (99% and more), and low hydrodynamic resistance of electrostatic filters. [6].

Let us consider a scheme for precipitating of contaminant particles in an electrical field created by charged fibers in an electret FM (for more detail, see Chap. 7). The operating principle of a fibrous electret filter is the interception of contaminants in response to coulombic and inductive forces (Fig. 6.5). The former are efficient in charged particle capture; the latter cause attraction of neutral particles by inducing dipolar electrical moments in them. Because the interceptive mechanism is conditioned by long-range electrical forces, an electret FM might have voids whose cross sections exceed the typical size of filtered particles. This is why electret filters display low pressure differentials at high filtration efficiency. For example, in filtering a submicron NaCl aerosol through a polypropylene electret filter at a flow rate of 20 cm/s, the pressure differential is 4 to 20 times less than that in a glass-fiber filter of similar permeability [9].

Electric forces are effective when either a FM or dust particles are charged. Dust particles or FM fibers can be charged by triboelectrification during filtration. Synthetic fibers acquire electret properties during molding and stretching in the melt blowing process (see Sect. 5.2). Fibers or contaminant particles can also be charged by, e.g., corona discharge (see Chap. 7).

When describing particle charging in a corona discharge field, the majority of scientists [10–12] proceed from the existence of two charging mechanisms, that is, ion and diffusion. The former includes particle charging by ions moving in an external electrical field; the latter is conditioned by diffusion of so

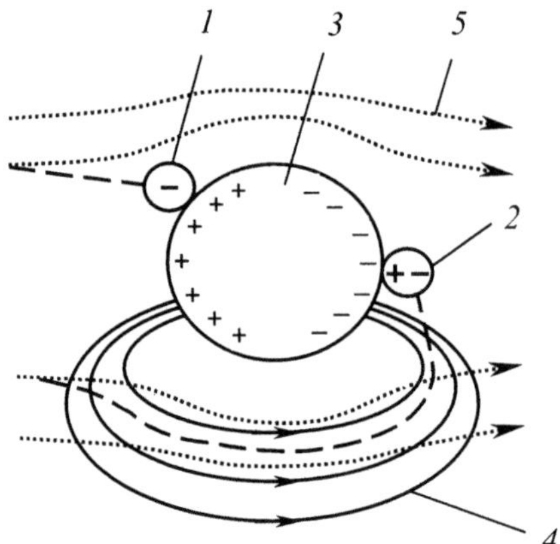

Fig. 6.5. A scheme of particle (*1* and *2*) attraction by a charged fiber (*3*) under coulombic (*1*) and inductive (*2*) forces; *4*, lines of electrical field intensity around a charged fiber; *5*, filtered aerosol flow lines

called thermal ions whose velocity depends on heat energy. Electrical field action is considered to charge particles mainly larger than 0.5 µm, whereas the diffusion process includes particles below 0.5 µm [6].

When moving in a medium a particle in an electrostatic field is affected mainly by electrical forces and gas-dynamic resistance. Using Coulomb's law to describe the former and Stokes–Kaningham's law for the latter, White obtained an equation to calculate the velocity of particle drift in a laminar gas stream [12]:

$$v_0 = \frac{qE}{6\pi\eta r}\left(1 + A\frac{\lambda}{r}\right) , \tag{6.1}$$

where q is particle charge, C; E is field intensity, V/m; η is gas viscosity, Pa·s; r is particle radius, m; λ is the mean free path length of surrounding gas molecules, m; and A is a dimensionless parameter (for air, $A = 0.86$).

Commonly, particle drift velocity noticeably determines dust collection efficacy. Equation (6.1), however, does not consider a number of factors arising from the nature of the dust, particle size inhomogeneity, and specific electrical resistance. These factors are taken into account by the notion of "effective drift (migration) velocity". Frederick [13] divided dust particles into three groups depending on their specific resistance. *The first group* includes particles with low specific resistance (below 10^2 ohm·m) for which dust layer discharge time is rather small. Therefore, a particle captured by a FE can eject back into the gas flow if its momentum is recharged. *The second group* involves dust particles with specific resistance of about 10^2–10^8 ohm·m. The particle discharge time is optimum for dust layer formation of sufficient thickness on the electrode or FE. *The third group* includes dust particles with high specific resistance (above 10^8 ohm·m). Such particles are difficult to intercept because the layer formed on the electrode (for example, on fiber) behaves like an insulator due to its considerable discharge time. This can bring about a so-called "reverse corona" and abrupt reduction of cleaning effectiveness of the gas flow.

Zimon [14] somewhat modified Frederick's classification by taking into account adhesive forces in dust deposition. Electrically active dust whose adhesive forces upon precipitation exceed cohesive forces and where particle aggregation does not take place are related to *the first class*. It includes burnt silicate and zinc oxide, cement, furnace black, corn starch, lumpy clay, and diatomite upon thermoalkaline treatment. *The second class* consists of electrically active dusts, which can discharge upon adhesion and form strong aggregates. It embraces smoke effluents of the zinc oxide type, raw cement, emission of a nickel melting furnace, magnesites, chromium ore, wheat starch, powdered sugar, and molybdenum oxide. To *the third class* belong electrically inactive dusts insensitive to the electrical effect of a FE, including soot, silicon dioxide, kaolin clay, and ground oats.

This subdivision is conventional because contact potential difference and charge leakage rate depend on dust properties and also on the properties of the filtering material as well. Thus, for example, fabric and fibers employed as filtering materials can be placed in a triboelectrical series depending on their ability to electrify from friction [15]. The triboelectrical series should be also treated as a rough estimate of filtering materials because the technological peculiarities of production and fiber surface modification can change the position of the material in the series.

6.2.4 Precipitation and Coagulation in a Magnetic Field

Using a magnetic field for cleaning liquid and gaseous media from impurities is, at first glance, limited by the specificity of the particles, that is, they must display ferromagnetic properties. First and foremost, these are iron-containing impurities formed from continuous or progressive corrosion and wear of machine parts. Note, that iron-containing particles also fulfill a supplementary function of transporting other contaminants and ions in the process of magnetic precipitation. This leads to deeper purification of liquids and gases even from the contaminants that do not precipitate in a magnetic field [16].

Magnetic precipitation is, as a rule, carried out by magnetized elements (packing) directly contacting liquid or gas flow. Iron-nickel spheres, shot, wire chips, crushed ferrite, ferromagnetic powder, ferrite magneto-soft granules, and magnetic fibrous materials can be used as magnetic elements of filters.

The probability of capturing particles with a magnetized FE is conditioned by the competing effects of different forces: inertial, magnetic, hydrodynamic (in the Stokes approach), buoyant (Archimedean), and gravitational.

If we consider only apparently dominating forces (magnetic, Stokes, and inertial), then the equation for the motion of a particle of mass m_0 is [16]

$$m_0 \frac{d\mathbf{v}_0}{dt} + \mathbf{F}_m + \mathbf{F}_s = 0 \,, \tag{6.2}$$

where $m_0 d\mathbf{v}_0/dt = \mathbf{F}_i$ is the inertial force; $\mathbf{F}_m = \mu_0 \chi V_0 \mathbf{H} grad H$ is the magnetic force; $\mathbf{F}_s = 3\pi\delta\eta(\mathbf{v} - \mathbf{v}_0)k_0$ is the Stokes force; $\mu_0 = 4\pi \cdot 10^{-7}$ H/m is the magnetic constant; χ, V_0, and δ are, respectively, the magnetic susceptibility, volume, and equivalent effective diameter of a particle; $\mathbf{H}$ is the magnetic field intensity in vicinity; η is the dynamic viscosity of liquid (gas); $\mathbf{v}$ and $\mathbf{v}_0$ are the liquid (gas) and particle velocity; k_0 is the dynamic particle shape factor (it is, on average, equal to 1.15 for a length to width ratio of particle size within 0.25–4).

When purifying from highly dispersed particles with low inertial forces, Eq. (6.2) will be capture condition in a scalar form:

$$F_m \geq F_s \,. \tag{6.3}$$

The correctness of this condition and the resultant regularities of magnetic filtration are supported by experimental data.

It has been found that magnetic force F_m depends to a great degree on particle size and its magnetic properties (magnetic susceptibility). In addition, particle precipitation intensity in an inhomogeneous magnetic field increases with increasing H and $grad\, H$.

If we imagine that a filtered medium is a flow of roughly similar particles in a carrying liquid (gas) and a magnetized filtering element is a quasi-uniform absorbing screen, then an equation can be written for magnetic filtering in a most general form:

$$\psi/\lambda = 1 - \exp(-\alpha L); \xi = \alpha L \,, \tag{6.4}$$

where $\psi = \frac{C_0 - C}{C_0}$ is a precipitation index of the active fraction of particles, C_0 and C are the initial and final content of ferromagnetic particles in the purified medium, λ is the portion of the active fraction of particles capable of magnetic precipitation, α is absorptivity, L is the filtering layer thickness, and $\xi = -1\mathrm{n}(1 - \psi/\lambda)$ is the logarithmic purification index.

Absorptivity α depends on magnetic field intensity H: $\alpha \sim H^f$, where f varies from 0.4 to 1. Consequently, the dependences of the precipitation index and purification factor on field intensity will acquire, respectively, logarithmic and exponential behavior. It is seen from Fig. 6.6. that the ξ versus H dependence is really close to exponential and exponent f can vary from 0.7 to 0.8 (for real contaminants in technological liquid and gaseous media, the mean value is $f \sim 0.75$).

Coagulation means sticking of colloidal particles together from directed motion during filtration under a force field action. It has been shown in the literature that particle precipitation in a magnetic field is attained due to

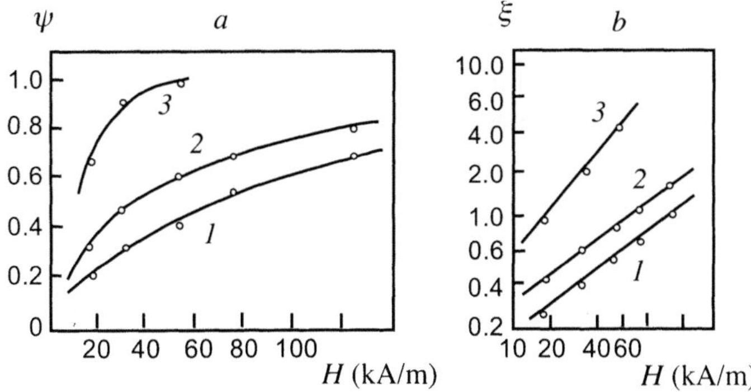

Fig. 6.6. The dependence of the precipitation index ψ (**a**) and logarithmic purification factor ξ (**b**) of magnetite particles on magnetic field intensity. The effective diameter of particles varies from tenths to tens of micrometers (1), below 8 μm (2), above 8 μm (3). The filtration rate $v = 5.6$ cm/s, and the magnetic filtering layer thickness $L = 4.2$ cm. The filtered liquid is an aqueous suspension of contaminants from a nuclear power station drain condensate [16]

intense *magnetic coagulation* in the dust-gas flow that leads to the formation of aggregates from ferromagnetic particles oriented along the field force lines. The aggregates reach 10–50 mm in length, and, what is more, when gases or liquids are purified by magnetic filters, coagulation and interceptive phenomena are observed with magnetic and also with other contaminant particles under the magnetic field effect. As a result, cleaning efficiency improves significantly. These problems are discussed in more detail in Chap. 8 using fibrous FM as an example.

Magnetic precipitation of particles can also be used for dusts that do not exhibit magnetic properties but have a preliminary charge. In a magnetic field of intensity H, an electrically charged solid particle moving in a gas flow at speed v_0 will experience a Lorentz force setting it into a spiral rotation. Within the Stokes law region the final drift velocity of a particle in electromagnetic field is determined by its charge, mass, and size, and is proportional to the linear velocity of the gas stream [6].

During gas flow filtration particle coagulation can take place in a magnetic and also in an acoustic field as well. *Acoustic coagulation* (agglomeration, flocculation depends on a number of factors, including sound radiation pressure, the probability of particle and gas joint oscillation (orthokinetic coagulation), and the relation between hydrodynamic forces of particle attraction and repulsion. The main reason for particle coagulation at frequencies up to 50 kHz is orthokinetic flocculation; at higher frequencies, the effect of air-dynamic forces dominates. In [11], the relations are illustrated for the critical frequency of sound waves, the critical size of particles isolated from the flow, the sound radiation pressure of oscillating gas, the particle velocity under this pressure, and the time of particle occurrence in an acoustic field. Optimum value regions for acoustic coagulation and particle precipitation are the following: sound oscillation frequency – 1–4 kHz; oscillation intensity – not less than 11.5 Wt/m^2; particle size – 1–10 μm; particle concentration – 2–200 g/m^3; time of particle contact in an acoustic field – 2–4 s.

7. Electret Filtering PFM

Cleaning of air and gas from suspended solid and liquid particles is of paramount importance in medicine, biochemistry, electronics, the atomic power industry and many other fields. The most simple, reliable and economical way of cleaning gaseous media of highly dispersed aerosols is using filters with a fibrous FE. The search for highly efficient purification systems able to remove submicron particles from air has led to the development of electret filters consisting of a FE with charged polymer fibers [1–4].

7.1 Mechanism of PFM Polarization

Because electret polymer fibrous materials (PFM) have extensively widening fields of application, it seems significant to improve such parameters as the magnitude and stability of the electret charge that affects the service characteristics of fibrous filters. In [5], results are cited of PFM polarization in a corona discharge, particularly, the effect of fibrous structure on the processes of electret charge generation and relaxation.

Test samples were 2 mm thick PFM plates obtained by melt blowing of LDPE. Fiber diameter did not exceed 50–70 μm, volume porosity of the material was about 50%. To estimate the polarization features of fibrous materials, reference samples were used as 400–500 μm thick coatings on aluminum foil produced by hot pressing initial PFM. The samples were polarized in a corona discharge at field intensity $E_\mathrm{p} = 3$ kV/cm, temperature $T_\mathrm{p} = 30$–$100°$C for $t = 10$ min.

In Fig. 7.1, the temperature dependencies of polarizing state parameters of melt-blown samples are compared with those of continuous coatings. Fibrous materials having a heterogeneous macro- and complex microstructure and greater in contrast to the coating thickness show lower j values than coatings. However, the T_p rise leads to current density increases in both cases due to the increased conductivity of materials (Fig. 7.1a).

Still more noticeable differences are observed for ESCD dependencies on polarization temperature (Fig. 7.1b). The electret coating shows a homocharge within the whole temperature range, i.e., the charge sign does not change (curve 2). The PFM electret under low T_p displays a homocharge, and at high T_p it acquires a heterocharge. The temperature at which charge

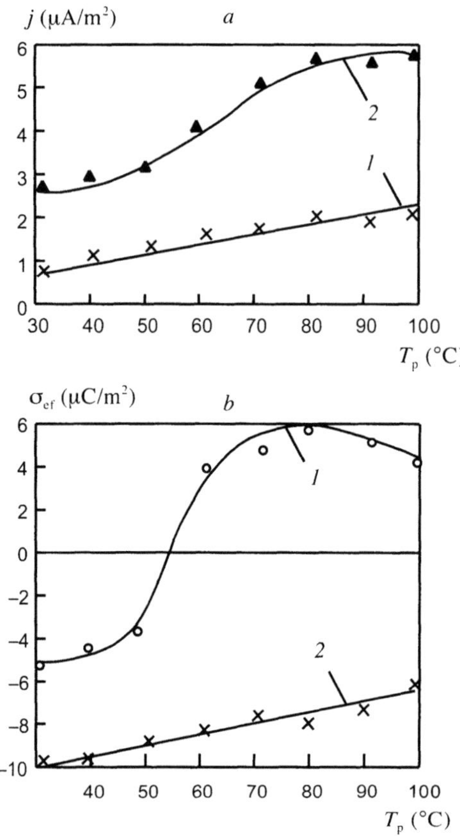

Fig. 7.1. The dependence of polarization current density (**a**) and ESCD (**b**) of PFM (1) and coating (2) versus polarization temperature T_p

inversion of ESCD takes place ($\sim 55°$C) is close to that of the λ_1 transition in polyethylene [6] which is related to "defrosting" of the molecular mobility of molecular net nodes of the noncrystalline phase. The T_p rise from 40 to $100°$C is accompanied by growing ESCD relaxation time in an electret PFM from 22 to 132 hours.

Information on electret charge generation in PFM can be obtained from the analysis of TSC spectra (Fig. 7.2). TSC spectra of coatings and PFM polarized at $T_\mathrm{p} = 30°$C (curves *1* and *3*) are characterized by one peak (maximum). Shift of the temperature corresponding to the PFM peak to the side of higher temperatures occurs because of the larger thickness of the samples and, consequently, their lower heat conductance compared to coatings. The TSC spectrum of PFM electrets obtained at $T_\mathrm{p} = 90°$C shows two inverse peaks (curve *2*). The first peak seen at $T_\mathrm{p} = 90°$C is related to homocharge relaxation, and the second is at $T_\mathrm{p} = 100°$C, which corresponds to hete-

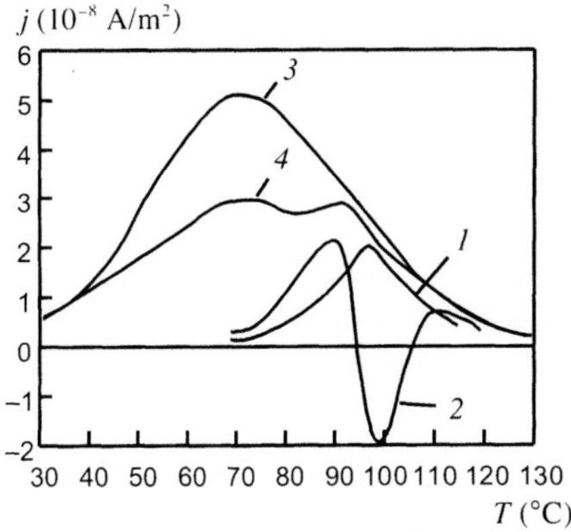

Fig. 7.2. TSC of LDPE corona electrets: *1* and *2*, PFM; *3* and *4*, coatings. $T_\mathrm{p} = 30°\mathrm{C}$ (1, 3); $T_\mathrm{p} = 90°\mathrm{C}$ (2, 4)

rocharge relaxation. The high-temperature peak of both coatings and PFM is more evident on the TSC of electrets polarized under high temperatures (curves *2* and *4*).

The appearance of two peaks on TSC spectra at high T_p provides the basis for the supposition that carriers entrapped at various depth within a forbidden band generate an electret charge in LDPE. The use of partial thermal cleaning of peaks [7] has made it possible to evaluate the activation energy W, the capture sections S, the escape attempt frequency of trapping centers ν_o, the residual charge value Q_res, and the charge distribution by types of trapping centers $M = Q_\mathrm{peak}/Q_\mathrm{gen}$ (Table 7.1).

As noted earlier, trapping centers in solid bodies can be subdivided into three types: coulombic centers of attraction, neutral, and coulombic centers of repulsion [8].

Trap cross sections vary from $10^{-15}\,\mathrm{m}^2$ (for centers of attraction) to $10^{-25}\,\mathrm{m}^2$ (for centers of repulsion). Strong capture centers show escape attempt frequency within the 10^8–$10^9\,\mathrm{s}^{-1}$ range, and those of a weak capture

Table 7.1. Localized state parameters of LDPE-based corona electrets polarized at $T_\mathrm{p} = 90°\mathrm{C}$

Sample type	No. of TSD curve peaks in Fig. 7.2	$T_\mathrm{max}(°\mathrm{C})$	W (eV)	ν_o (s^{-1})	S (m^2)	Q_res (nC)	M (%)
Coating	1	73	0.80	$1.7 \cdot 10^9$	10^{-21}	1.10	70
	2	92	1.14	$5 \cdot 10^{13}$	10^{-16}	0.47	30
PFM	1	89	1.20	$6 \cdot 10^{14}$	10^{-15}	0.15	31
	2	100	0.90	$6 \cdot 10^8$	10^{-20}	0.33	69

are 10^{13}–10^{14} s^{-1}. Analysis of Table 7.1 allows us to draw the conclusion that charge carriers in both PFM and coatings localize in two types of traps, that is, neutral centers that realize strong capture (peak 1 for coatings and peak 2 for PFM) and coulombic centers of attraction that realize weak capture (peak 2 for coatings and peak 1 for PFM). Major portion of electret charge is formed by carriers localized in neutral trapping centers as in coatings, so in PFM. For PFM, it is typically carriers that generate a heterocharge.

The analysis of experimental data makes it possible to suggest the following mechanism for homo- and heterocharge formation in electret fibrous material.

Under low-temperature polarization of LDPE (within the 30–50°C range), filling of localized states, both centers of attraction and neutral, proceeds owing to carriers injected by corona discharge. Entrapped by the centers, carriers generate a homocharge in both coatings and PFM. Their delocalization during thermal depolarization leads to formation of only one TSC peak (Fig. 7.2, curves 1, 3).

The increase of polarization temperature ($T_\mathrm{p} = 60$–$100°$C) leads to augmented equilibrium conductivity of the material, i.e., to a heat-induced increase of carrier concentration within the polymer conduction band. In a continuous coating, intrinsic carriers drift to the corona electrode and recombine with carriers injected by corona discharge. Because the concentration of intrinsic carriers is lower than that of injected carriers, the portion of unrecombined injected carriers localizes in the trapping centers to form a homocharge. Its value is, nevertheless, less than that of the charge formed in low-temperature polarization. This can be seen from the reduction of the area beneath curve 4 in contrast to that of curve 3 (Fig. 7.2).

In polarizing PFM with a discrete structure, the migration of intrinsic and injected charge carriers is complicated. As a result, intrinsic carriers localize in neutral trapping centers within the fibrous bulk at the interface between crystalline and noncrystalline phases and thus create Maxwell-Wagner polarization. These carriers form a heterocharge, and their delocalization in thermal depolarization gives rise to the second (negative) TSC peak (Fig. 7.2, curve 2). Its location in a high-temperature region is connected with inertial heating of the fibrous material that have low heat conductance. Corona injected carriers localize in coulombic centers of attraction in fiber surface layers and form a homocharge which produces the first (positive) TSC peak in depolarization.

During PFM production by melt blowing, the polymer is subjected to intense temperature and mechanical effects leading to its thermal and thermally oxidizing destruction. The latter is intensified in polarization by corona discharge at elevated temperatures. These processes are accompanied by an increasing number of unsaturated links induced by formation of vinylidene, vinyl and *trans*-vinylene groups in the fiber bulk, as well as hydroxylic and carbonyl groups on its surface. The groups with easily polarizable double links (vinylidene, vinyl and *trans*-vinylene) interact with charge carriers (intrinsic or injected) as neutral trapping centers making capture, probably, analogous to the mechanism of small-radius polaron formation [8]. Charge carrier trapping by the centers of attraction occurs, probably, due to their coulombic interaction with polar groups (hydroxylic and carbonylic) concentrated mainly in the fiber surface layers.

Carrier localization in neutral trapping centers is advantageous from the viewpoint of increasing the temporary stability of the electret charge. This is explained by the absence of the Pool–Frenkel effect [9] in these centers, i.e., no pótential barrier reduction that hampers charge carrier delocalization is observed in the strong inner field of the electret.

From the analysis of probable polarization mechanisms of fibrous materials by corona discharge, the following conclusions can be drawn:

- Heterocharge formation in PFM takes place in polarization at elevated temperatures, is conditioned by a nonhomogeneous material structure, and follows a mechanism analogous to that of Maxwell–Wagner polarization in heterogeneous dielectric media.
- Electret heterocharge carriers are not injected but are intrinsic and are localized in neutral trapping centers which assists in increasing the electret charge relaxation time and the electret stability.
- It can be anticipated that high-temperature treatment of PFM by corona discharge along with spontaneous fiber polarization during melt blowing (see Sect. 5.2) makes it possible to achieve electret charge parameters that will provide highly efficient trapping of submicron size particles by a fibrous polymer FE.

7.2 Capillary Phenomena

The PFM structure presupposes the presence of a developed system of pores and capillaries in the filtering material bulk. It is common knowledge [10] that high relative air humidity gives rise to *capillary condensation*. This phenomenon can significantly effect the gas filtration process. The liquid layer between a contaminant particle and the fiber surface in electret filters substantially weakens the effect of electrical forces. In an air medium with high humidity, capillary forces prevail over other forces responsible for retaining particles on a fiber surface. Capillary forces depend on the physicochemical properties of interacting surfaces, but mainly on the hydrophilic-hydrophobic balance.

Capillary condensation can be presented schematically as follows [10]. In the clearance between contacting bodies, water vapor condenses (Fig. 7.3). The meniscus formed draws the particles together by surface tension forces, on one hand, and reduces capillary liquid pressure due to its concavity (in conditions of surface wetting), on the other hand. In full wetting, $\gamma \to 0$

(Fig. 7.3), and the formula for capillary interaction between two spherical particles is

$$F_{k1} = 2\pi\sigma_{sl}r \ , \tag{7.1}$$

where σ_{sl} is the coefficient of surface tension at the solid-liquid interface, and r is the particle radius. If one of contacting surfaces is flat (Fig. 7.3b,c), then the liquid layer thickness h becomes approximately two times less; then

$$F_{k2} = 4\pi\sigma_{sl}r \ . \tag{7.2}$$

Relations (7.1) and (7.2) are true in conditions of full wetting of smooth surfaces ($\gamma \to 0$) and for relatively large particles. For nonwettable (or not fully wettable) surfaces, the right-hand sides of (7.1) and (7.2) should be multiplied by $\cos\theta$ (θ is the edge wetting angle); then

$$F_{k1} = 2\pi\sigma_{sl}r\cos\theta \ , \tag{7.3}$$

$$F_{k2} = 4\pi\sigma_{sl}r\cos\theta \ . \tag{7.4}$$

According to (7.4), capillary forces exert the most influence on particle adhesion to a fiber when the surfaces are hydrophilic ($\theta \to 0$) and the least effect when they are hydrophobic ($\theta \to 90°$).

At 100% relative humidity when particle sticking is caused exclusively by capillary phenomena, experimentally estimated adhesive forces [10] are less

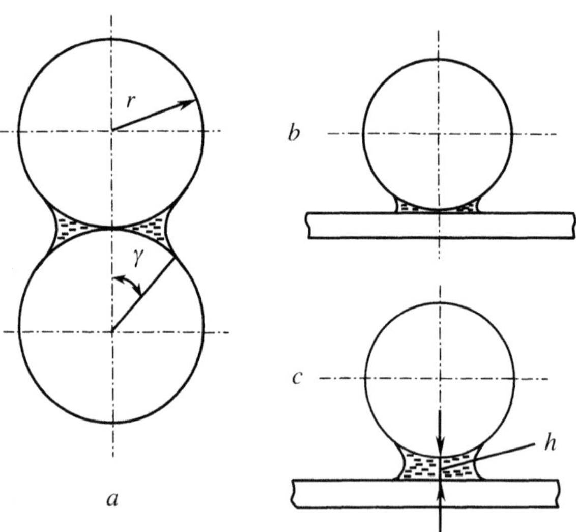

Fig. 7.3. Diagram of capillary condensation in the contact between two spherical particles (**a**) and of a particle with a fiber (**b**, **c**): (**a**), (**b**), without a liquid layer; (**c**), with a liquid layer in the contact area

than those calculated by formulas (7.2) and (7.4). The point is that these equations do not consider the wedge pressure of the thin liquid layer, which weakens capillary interaction.

Wedge pressure characterizes the intensity of interaction between surface layers of a thin liquid film in the clearance between contacting bodies. It is determined by the difference between the thermodynamic and chemical potentials of the layers from those in the bulk liquid phase due to the influence of neighboring (bordering on the film) phases. The notion of wedge (disjoining) pressure for a thin liquid layer between the surfaces of solids was established by B.V. Derjaguin [11,12] and was expanded afterward to a wide circle of problems, including the stability of lyophobic colloids [13], lubricity [14], moisture equilibrium, motion in the ground, and other processes [15].

Wedge pressure is conditioned by two phenomena: molecular (van der Waals) interaction between a solid phase and a boundary layer of the liquid (molecular component) and formation of an electrical double layer at the interface (electrical component). The electrical component includes, in its turn, ionic and diffusion constituents. So, it turns out that the total wedge pressure of a liquid layer of thickness h is

$$P(h) = P_e(h) + P_m(h) = P_i(h) + P_d(h) + P_m(h) \, , \tag{7.5}$$

where indexes e, i, d, and m are, respectively, the electrical, ionic, diffusive, and molecular constituents of wedge pressure.

The electrical component of wedge pressure is related to the electrical field intensity E which is directed normally to the contact surfaces of the bodies. The Derjaguin–Landau theory gives an opportunity to evaluate the electrical component of wedge pressure for low-concentration electrolytes, e.g., condensed liquid films on charged fiber surfaces in an electret FE [13]:

$$P_e(h) = \frac{\varepsilon E^2}{8\pi} = \frac{\varepsilon}{8\pi} \left(\frac{\partial \varphi}{\partial h} \right)^2 \, , \tag{7.6}$$

where ε is the dielectric constant of the condensate, and φ is the fiber surface potential. As calculations have shown [16], wedge pressure depends on the clearance h between contacting bodies and increases with a reduction in h.

Another factor influencing capillary force interaction between a particle and a fiber (see formula 7.4) is variation in the edge wetting angle θ under the action of an electric field created by the charged fiber (the so-called electrocapillary effect).

The electrocapillary phenomenon, first studied by A.N. Frumkin, is the variation in surface tension at the solid-liquid interface owing to the electrical potential jump at the interface in metal electrode-electrolyte systems [17].

Data are available on both improving and impairing the wettability of polymer materials subjected to electrification. For example, thermal [18], corona [19], and radio-electret [20] polymer films and coatings are more water-wettable than corresponding unpolarized samples. Independent of the surface charge sign, edge water-wetting angles of HDPE and PTPCE-based thermo-electrets reduce with increasing ESCD [18]. Improved wettability in these cases is attributed to increased surface energy at the polymer-air interface in polymer polarizing. At the same time, it has been found that edge glycerin-wetting angles of thermal and corona-electrets made of pentaplast [21] and some minerals subjected to γ radiation [22] increase substantially. These results, at first sight contradictory, can be explained as follows.

According to adopted notions on wetting processes, the value of the equilibrium wetting angle depends on the correlation between surface tension forces active in wetting at the interface (solid body, liquid, gas). This dependence is expressed through Young's law,

$$\cos\theta = (\sigma_{sg} - \sigma_{sl})/\sigma_{lg} , \tag{7.7}$$

where σ_{sg} and σ_{lg} are the coefficients of surface tension, respectively, at the solid-gas and liquid-gas interfaces. Relation (7.7) can be presented as the Dupret-Young equation relating the work of adhesion W_a to the liquid surface tension and the edge angle:

$$W_a/\sigma_{lg} = 1 + \cos\theta , \tag{7.8}$$

or as the correlation between the work of adhesion and cohesion:

$$W_a/W_c = (1 + \cos\theta)/2 , \tag{7.9}$$

where $W_a = \sigma_{lg} + \sigma_{sg} - \sigma_{sl}$ is the work of adhesion, and $W_c = 2\sigma_{lg}$ is the work of cohesion.

Analysis of Eq. (7.7)–(7.9) shows that wetting quantitatively determined by $\cos\theta$ depends on the correlation between the adhesive and cohesive work of a wetting liquid. A liquid, whose surface tension σ_{lg} or cohesive work is less, wets better.

To comprehend the regularities of the effect of solid body polarization on wetting and spreading, it is important to take into account that the moving force of spreading $\Delta\sigma$ per unit length of wetting perimeter is conditioned by those constituents of the system's free energy which are diminishing by wetting [22]:

$$\Delta\sigma = \sigma_{sg} - \sigma_{sl} - \sigma_{lg}\cos\theta_d , \tag{7.10}$$

where θ_d is the dynamic edge wetting angle.

Summarizing, it can be supposed that solid body polarization leads to increasing adhesive work by increasing its free surface energy, i.e. it assists wetting and spreading. On the other hand, an electrified substrate brings about polarization of a contacting liquid and increases the cohesive forces inside a droplet. Liquid contact polarization reduces the moving force of spreading, $\Delta\sigma$. Depending on what effect prevails, both wetting improvement and spreading acceleration, or vice versa, can be observed as a result of polarization.

For viscous easily polarizable liquids, e.g., glycerin and DEG studied in [21], their contact polarization effect probably prevails. It is also not excluded that wetting of electret polymer films by glycerin or DEG is hampered by liquid diffusing and dissolving in the polymer surface layer. The more intensive liquid transfers through the interface, the stronger the σ_{sl} reduction leading, in its turn, to increased W_a and wetting relief. As proved in [23], polarization of a polymer film retards liquid diffusion transfer through the interface and, consequently, hampers wetting.

Thus, understanding the polarizing charge effect in liquid wetting and spreading over a polymer electret surface provides a basis for forecasting the efficiency of an electret PFM in air filtering systems where relative humidity is high, as well as in gas cleaning where organic liquid vapor content is high. The efficiency of an electret fibrous FE based on PE is assumed to be impaired in air filtration in elevated humidity because polarized fibers are wetted better by water whose drops spread perfectly across the fiber surface, thus reducing the electrostatic interaction between the fibers and dust particles. In this case, the retention of dust particles on the fiber surface is determined by capillary force acting through the layer of wetting liquid. A different situation is observed in oil mist filtering through an electret FE based on PP. Oil drops deposit on fibers without wetting and are retained there by electrostatic forces. In this case, the filter efficiency will be higher if fibers possess an electret charge.

7.3 Production Process and Properties of Electret PFM

Mainly PFM with preliminarily charged fibers are employed for manufacturing electret FEs. The fibers are charged in a high-voltage field or in a corona discharge.

A method is described in [4] for producing an electret FM from a polypropylene film stretched to increase its mechanical strength and subsequent electrification in a corona discharge. Preliminary unidirectional extension of the film makes its splitting into separate fibers easier after charging. The film can be charged one-side or two-sides in corona discharge. In the first case, the film is bipolar, i.e., it attains a negative charge on one side and a positive charge on the other. The opposite charge is generated in the air gap between the film and the metal plate along which the film slides during charging. The method of two-side charging by corona devices that inject opposite sign charges makes it possible to increase film speed in charging to 100 m/min and to increase the charging stability owing to reduced losses from spark breakdown.

The production of electret FMs from the films has not been commercially adopted because of, first, laboriousness and low productivity and, second, the physicomechanical characteristics of such a PFM are not high enough. The latter disadvantage has been partially eliminated in Petryanov's FM produced by a solution technique [24,25] that is employed in highly efficient dust-fighting respirators. The present-day manufacturing of fibrous filters is characterized by a tendency to supplant ecologically harmful technologies, therefore, melt-blown PFM successfully compete with Petryanov's FM. The high productivity of the melt blowing method and the wide capability of controlling the fiber charge state allows us to forecast the future production growth of highly efficient PFM-based electret filters.

Use of the PFM polarization technique during polyethylene melt blowing has been patented [26]. The fibers are deposited on a revolving steel drum in a high-voltage electrical field or corona discharge. To create a high-voltage electrical field a system of coaxial cylindrical electrodes is used, one of which is the grounded receiving drum, and the high-voltage electrode embraces the drum brass cylinder with a strip recess along the generatrix. The cylinder is installed so that fiber spraying is carried on through the recess on the drum surface.

Corona discharge is created by a system of needle electrodes 1 mm in diameter installed in the cylinder at a 20 mm pitch and 10 mm from the drum surface. A negative potential is applied to the corona electrodes from a high-voltage source.

In Table 7.2, the main characteristics of electret FM on a PE base are shown with fiber diameter of about 100 μm, depending on the PFM electrification regimes. Filtering parameters were evaluated by the degree of purification of a solid aerosol with particles 0.5–0.6 μm in effective size. In contrast, a FE obtained by a similar melt-blowing technique but without electrical fields displays approximately two times higher resistance, and skipping coefficient $K = 1.1$

Table 7.2. Characteristics of electret PFM [26]

Method of fiber electrification	Field intensity in spraying zone (kV/cm)	Material characteristics			
		ESCD[a] $\sigma_{\rm ef}({\rm nC/cm}^2)$	Density $\rho\,({\rm g/cm}^3)$	Pressure drop Δp across filter (Pa)	Skipping coefficient $K\,(\%)$
In high-	5	+0.9/−0.4	0.25	10.1	0.30
voltage field	25	+1.9/−1.0	0.30	10.2	0.10
	50	+2.2/−1.5	0.25	10.0	0.10
In corona	5	−2.3/−1.5	0.20	10.3	0.10
discharge	10	−2.0/−0.9	0.19	10.1	0.10
	15	−1.5/−1.0	0.22	10.5	0.19

[a] In numerator and denominator – on opposite sides of the sample.

The most rigid requirements for aerodynamic resistance and skipping coefficient are imposed on fine-cleaning filters. The antismoke FEs of gas masks, respirators, etc. are tested in a standard oil mist with a mean droplet size of 0.1–0.2 μm and a filtration rate of about 1 cm/s.

Electret FMs obtained from melt-blown fibers based on LDPE and PP, subjected to corona discharge, show the following characteristics: fiber diameter 1–5 μm, packing density 0.37–0.55; $\sigma_{ef} = 11$–15 nC/cm^2. Such 3–4 mm thick FEs ensure a skipping coefficient of less than 0.1standard oil mist and resistance to air flow below 40 Pa [26].

In prolonged storage under mechanical loading, ionizing radiation, or high humidity, leakage of electret PFM charges occurs. Discharge of an electret FE is also the result of an accumulation of conducting dust. Filter efficiency drops under high filtration rates (above 20 cm/s) because traveling particles do not succeed in sticking to fibers under electrical forces. Nevertheless, available data show that electret PFM preserve a sufficiently stable charge at elevated temperatures and humidity, as well as when dipped into water or alcohol. FE testing at 80°C for 100 days leads to impaired filtration efficiency from 99.5 to 92% [4].

The choice of polymer melt-blown materials for electret filters is restricted by the chemical and thermal resistance of the polymer, its ability to gain a stable charge in time, and its processability into fibers with acceptable mechanical characteristics. If PFM are produced from polymer blends, their electret charge can be increased. In [28], it is shown that the ESCD value in corona electrets based on a mixture of HDPE (75 mass%) and PVC (25 mass%) exceeds the analogous index of a corona electret based on pure HDPE by 2.2–2.4 times. This is because the electret charge of polymer blends electrified by corona discharge is mainly induced by ionization of localized energy states formed at phase interfaces. Polyethylene possesses a more heterogeneous structure compared to PVC because of numerous interfaces of crystalline and noncrystalline phases. The structural heterogeneity of their blend characterized by interfacial areas of structural elements is greater in contrast to pure components. This produces a higher density of surface states and a higher capacity of blends to be electrified. A dominating role in electret charge generation and transformation in polymer blends is played by Maxwell–Wagner polarization, whereas volume-charge and dipolar polarization are typical for homogeneous materials.

The choice of fiber diameter and density of electret FM is predetermined by the degree of cleaning and the filtration regimes: rate, temperature, humidity of filtered gas flow, and filter service life. The melt blowing technique enables variation of fiber diameter and FM density in a wide range. Filters with a FE of variable density are more efficient. Thus, FE can be made of different density layers consisting of synthetic fibers of different diameters [29]. One of the blend components has, as a rule, a fiber diameter 4–10 times larger than the rest, and the amount of this component is varied following a specific law in the direction of the filtered gas stream. The cleaning efficiency of such filters is intensified by reduction in the turbulence of the filtered gas. Increasing aerodynamic resistance within the filter inlet-outlet path reduces the probability of particles tearing from the fiber surface.

7.4 Applications

The main fields where PFM-based electret filters are used are fine purification of air and personal means of protecting respiratory organs (respirators).

Fine air purification in industrial locations presumes the reduction of highly dispersed aerosol concentrations in air to values less than permissible limits or required in a process. This is especially required for the production of radio and electronic instruments, nuclear power, the manufacture of medicinal preparations, and superpure chemical substances.

Fine-purification filter (FPF) efficiency is determined by the aerodynamic resistance (or thickness) of the filtering layer and the air filtration rate. The thicker the FM layer, the higher the filter efficiency. However, FM thickening leads to increased filter resistance. PFM-based FPFs ensure the needed efficiency of purification systems at rather thin filtering layers. Besides high efficiency, FPFs should have sufficiently high mechanical strength, be a convenient size and shape, assist sealing, and create a few times less resistance to air flow than the pressure from the pumping device [24].

Most widely spread FPFs for air ventilation are the frame and hose designs. The frame filter consists of a box-type casing where the FE is installed in a specific manner. It is made as a strip of FM fixed tightly on the bottom and cover of the box. The strip is pleated to make up a set of frames in a U formopen from opposite sides of the box. Air is supplied to the casing and enters the frames open from the side of the inlet pipe. Then it is filtered by the FM strip and sucked off the frames open from the opposite side of the casing through the output pipe. The geometrical parameters of the frame depend on the filtration velocity and filter resistance, i.e. based on filter designation and required degree of air purification.

Hose type filters are made as cylinders 20–30 cm in diameter and up to several meters long. These filters have found much use owing to their simple design, low resistance to the filtered air stream, and easy substitution of a used FE.

Electret PFMs have made it possible to elaborate new types of *respirators* displaying high filtration efficiency and ease of handling.

The respirators are intended for protecting respiratory organs from metallurgical, ore, cement, coal, tobacco, tea, and other dusts contained in air in concentrations exceeding 200 permissible limits. The main element of the respirator is the housing that serves simultaneously as a filter and a mask and consists of one or more layers of filtering material. The skeleton of the housing is supported by a spacer or stiffening ribs installed on the inner layer of the mask made by hot pressing a fibrous material. The mask fits snugly to the face owing to a rubber lace or a strip of soft foam plastic around its inner perimeter. This practically excludes penetration of unfiltered air through the seal and makes respirator use suitable for people with any facial features.

I.V. Petryanov's filtering materials made of polymer solutions have long been used in Russia for respirator production. The material of Petryanov's filter (PF) is a system of isotropic layers of ultrathin ($\sim$ 0.3–1 µm) polymer fibers applied on a cloth substrate (gauze, percale, or coarse calico) or on a base of bonded thick polymer fibers. Despite their satisfactory service characteristics, the production process is rather costly and ecologically unsafe. That is why the problem of PF substituting by cheaper and more technologically effective filters with no less performance has now become actuality in CIS countries.

The filtering properties of electret melt-blown PFM based on polypropylene have been studied by authors [27]. The objects of investigation were three types of PFM sheet samples: I, single-layer, uncharged 3-mm thick sheet; II, the same thickness sample charged by corona discharge (10 kV/cm) during polymer melt blowing; III, two-layer PFM having an uncharged first layer ($\sim$ 1.5 mm) and charged during melt blowing by corona discharge second layer. Filtering efficiency was judged from the skipping coefficient K of an oil aerosol through PFM samples.

In Fig. 7.4a, the kinetic dependence of coefficient K shows the character of filter degradation during service life. As can be seen, K of nonelectrified sample I (curve *1*) ascends smoothly and stabilizes with time and then abruptly drops after the first hour of operation. The pores in the FM surface layer probably become blocked with aerosol particles. Particle capture can take the following pattern. Oil mist drops spread over the fiber surface and form a film that gains thickness as the aerosol arrives. Under the action of a gas stream and gravitational forces, the liquid piles up in contacts with the fibers forming droplets which first cut off the fine pores and then the coarser ones. A natural obstacle that hampers mass penetration of aerosol particles into the FM occurs that leads to elevated aerodynamic resistance of the sample [1].

The K value of a single-layer PFM of type II sample (curve *2*) rises negligibly for some time and then reaches a constant magnitude. This can be attributed to an insignificant decay of the electret charge due to charged fiber screening by an oil film and its stabilization at saturation of the charged fiber sorptive volume [21]. It corresponds to a state when aerosol particles (oil drops) are distributed evenly along fiber surfaces. The behavior of curve *3* (a two-layer sample, type III) is probably governed by a total effect arising from electrified and nonelectrified layers of the samples.

The K dependence on dispersed aerosol particle diameter is presented in Fig. 7.4b. The two-layer PFM has a lower K coefficient than type I and II

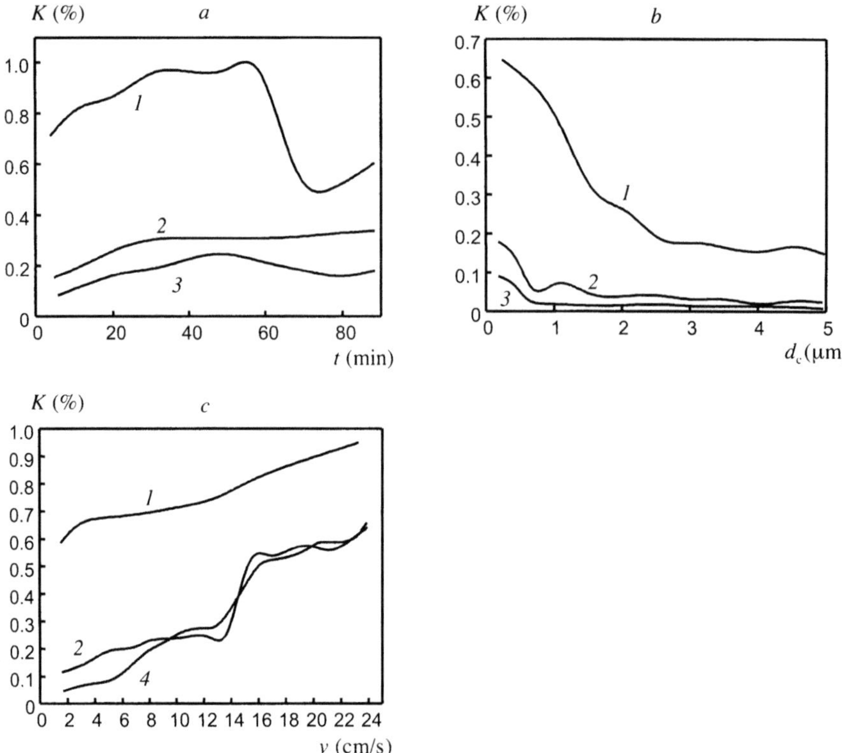

Fig. 7.4. The dependence of aerosol particle skipping coefficient versus (**a**) test time ($d_c = 0.2$ μm, $\nu = 1.5$ cm/s); (**b**) aerosol particle diameter ($t = 5$ min, $\nu = 1.5$ m/s); (**c**) filtration rate ($d_c = 0.2$ μm, $t = 5$ min); 1, nonelectrified sample (type I); 2, sample electrified by negative corona discharge (type II); 3, two-layer sample with charged and uncharged layers (type III), 4, sample (like type II) charged in positive corona discharge

samples, and its filtration efficiency in capturing particles with mean diameter $d_c = 3$–5 μm approaches 100%.

The filtration efficiency of electret PFM depends little on the corona discharge polarity used for electrification (Fig. 7.4c). The K of a positively charged PFM is somewhat lower than that of a negatively charged material at a filtration rate $\nu = 1$–7 cm/s of an oil aerosol, however, the values practically coincide when the velocity approaches 8–15 cm/s.

The reduction of filtration efficiency is accompanied by almost linear growth of aerodynamic resistance R of filtering materials with increasing filtration rate. The least resistance to air stream is typical for electret PFM that emphasize a more ordered distribution (under electrostatic field action) of captured particles across the fiber surface. A great volume of connecting vacancies remain unoccupied within the interfiber space, which is why the filtration efficiency of PFM does not drop. This supposition is supported by aerodynamic resistance two to four times higher in a nonelectrified FE that executes only mechanical capture of contaminants compared to an electret FE. Mass choking by aerosol particles probably takes place in a nonelectret FM that causes an R increase.

So, when PFM are used in fine gas filtration, including respirators, materials combining electret and nonelectret filtering layers have proved to be the most efficient. Alternation of these layers as well as thickness variation perceptibly improves PFM filtration characteristics.

The diversity of operating regimes has resulted in the development of a multitude of respirator modifications. Several types of "Lepestok" respirators are produced in Russia where PF materials are gradually substituted by melt-blown electret PFM. One of the major manufacturers of means for protecting respiratory organs, 3M Co., produces respirators with a multilayer FE (three to seven layers) where the middle layer carries the electret charge. These respirators possess insignificant resistance to breathing (1–20 Pa) and display 96 to 99.9% efficiency in filtering 0.1–0.2 µm particles. Their service life is rather long and lasts from one labor shift (in protection against highly toxic substances) to 8–10 (in low concentrations of usual dusts).

8. Magnetic Filtering PFM

The V.A. Belyi Metal-Polymer Research Institute of NASB (MPRI) has elaborated a new class of filtering materials – magnetic PFM [1–8]. The technological base for manufacturing such materials is the melt blowing technique involving the following procedures: extrusion of a polymer melt filled by ferrite (barium or strontium) powder, fiber extension by gas flow, and fiber treatment in a magnetic field. The polymer melt is extruded through spinneret holes whose diameter far exceeds that of the filler particle. The thermal regime of spraying provides for cohesive bonding of the fibers on the forming substrate. The material is also textured during spraying. The final stage of magnetic PFM or finished FE manufacture is filler particle magnetizing in a permanent or pulse magnetic field.

8.1 Background

A means of improving the properties of lubricating oils and working fluids in hydraulic systems is cleaning them from wear debris. Because the major part of the impurities, as a rule, possesses ferromagnetic properties, there are real grounds for applying methods and devices to precipitate these particles magnetically. Magnetic filters based on either permanent or electrical magnets are in a wide use at present [9]. So, particles from a filtered liquid are precipitated magnetically in a ferromagnetic FE located inside a hollow magnetizing coil or permanent magnet. The degree of filtration is a function of the magnetic field intensity, flow rate, viscosity of the working liquid, and the origin and size of the impurity particles. Filtration efficiency can be increased by increasing the FE mass or dimensions, although in many cases it is impossible or not justified from an economic viewpoint.

Analysis of filtration principles and design solutions employed in engineering have proved that fibrous FM generating magnetic fields are a most simple and efficient way of solving the problem of purifying liquids that contain iron impurities. In contrast to fibrous electret materials employed for many years to clean gases, no large customer magnetic has yet recognized fibrous FM. The reasons are high power consumption and the technological complexity of the production process for ferromagnetic fibrous materials.

The structure of magnetic PFM is a combination of cohesively bonded fibers with incorporated ferrite filler particles (Fig. 8.1). It is seen from the figure that "thick" (up to 100 μm in diameter) filled fibers alternate with thin ones that have encapsulated ferrite particles cohesively fixed on their surfaces. Mark "+" denotes the location of fiber section (b) whose results by XSMA were previously cited in Fig. 5.4.

By choosing a technological regime of magnetic PFM formation, their density, porosity and mechanical parameters can be controlled across a wide range (see Sect. 4.2). As an example, the range can be taken of the physicomechanical characteristics of PE-based PFM containing 15 mass% of barium ferrite: density, 0.2–0.6 g/cm^3, porosity, 30–80%, fiber diameter, 5–200 μm, breaking stress under tension, 0.3–5.0 MPa, and relative elongation at rupture, 60–180%.

The filtration efficiency of liquids containing ferromagnetic impurities is determined by a combination of a number of factors. It depends on such parameters of magnetic PFM as ferrite particle size and concentration, their magnetization level, FM density and fiber diameter, FE thickness, as well as on the characteristics of the filtered medium and the filtration conditions, contaminant particle size, and filtration rate etc.

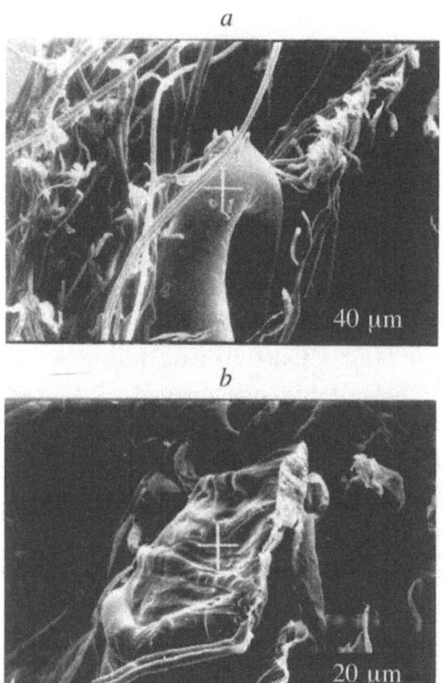

Fig. 8.1. Electron microphotographs of magnetic PFM based on PA (**a**) and PE (**b**) filled with 15 mass% of strontium ferrite

8.2 Simulation of Magnetic Deposition in PFM

The process of cleaning a liquid of fine (below 20 µm) contaminating particles using a magnetic FM has its specific features. In [10], the efficiencies of various filtration mechanisms in ferrite-filled PFM are estimated using theoretical dependencies for the precipitation energy of contaminant particles. The magnetic interaction energy E_{m} between fiber and particle exceeds the energy of van der Waals attraction E_{w} and diffusive precipitation E_{d}. The latter value is considerably less than (by a few orders) the first two and can be neglected. In this way, the main mechanisms affecting fine iron-containing impurities are interceptive and magnetic precipitation.

Calculation of E_{m} and E_{w} energies has shown that, even for nonmagnetized FM filled by single-domain ferrite particles where the distance between ferrite particle and contaminant $r \leq 2$ µm, energy E_{m} exceeds E_{w} roughly by three to seven times. Magnetization of the material augments the magnetic interaction energy in response to an increasing induced magnetic moment on contaminants and brines about their coagulation with more effective precipitation. Other conditions being equal, the increase in the amount of ferrite and its magnetization assist in intensified magnetic interaction between the FM and contaminants (Fig. 8.2). At the same degree of filling, more intensive magnetic interaction is realized in magnetized FM filled with single-domain ferrite particles.

The efficiency of contaminant precipitation has been evaluated on the simplest model of a magnetic FE. The model presents an individual fiber with spherical particles of magnetosolid ferrite filler evenly distributed inside it [7]. The model is based on a theory according to which filter productivity

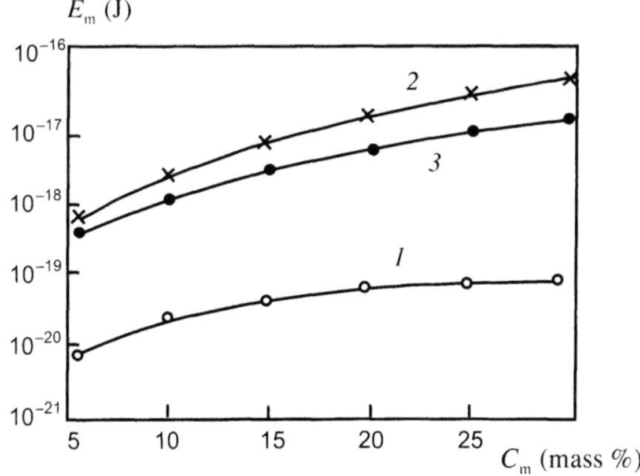

Fig. 8.2. Calculation dependencies of magnetic interaction energy between contaminant and ferrite particles in PFM: 1, magnetized FM filled by agglomerates of ferrite particles; 2 and 3, magnetized and nonmagnetized materials filled by single-domain ferrite particles

is ensured by the capability of an individual fiber of capturing contaminants from the fluid stream of a given viscosity. Proceeding from this, the efficiency of solid contaminant particle capture by a fiber is reflected by the relation [11]

$$E_f = \frac{r}{2r_\mathrm{f}La}\left[2\ln\left(\frac{r}{r_f}\right) - 1 + \left(\frac{r_f}{r}\right)^2 + \frac{F(r)\cdot\beta}{v}\right] ,\tag{8.1}$$

where r is the particle to fiber distance; r_f is the fiber radius; $La = 2.002 - 1\mathrm{n}Re$, where Re is the Reynolds number; $F(r)$ is the force of magnetic attraction between the particle and fiber; $\beta = 1/(6\pi r_\mathrm{c}\eta)$, where r_c is the contaminant particle radius, η is the liquid dynamic viscosity; and v is the mean liquid flow rate inside the filter (around the fiber). The addend $F(r)\beta/v$ in Eq. (7.1) characterizes the magnetic interaction between contaminant particles and the FM.

Let us consider four models of fibrous FE differing by the value $F(r)$: I, fibers without magnetic filler, $F(r) = 0$, and the last member of the right-hand side in relation (8.1) is absent; II, fibers contain magnetic filler as single-domain ferrite particles, and the FE is nonmagnetized; III, the same as in II, but the FE is magnetized; IV, fibers contain agglomerated ferrite particles and the material is magnetized. In Table 8.1, formulas are presented for models II–IV used to calculate the force of magnetic attraction between a contaminant particle and a fiber [12]. The filtration efficiency of a fibrous FE has been calculated according to Davies's equation [11]:

$$E = 1 - \exp\left(-\frac{2E_\mathrm{f}ch}{\pi r_\mathrm{f}}\right) ,\tag{8.2}$$

where c is the FM packing density (the ratio of FM apparent density to the density of material from which the FE is produced); and h is the FE thickness.

In Figs. 8.3a and 8.4a, calculation dependencies are shown for water and mineral oil filtration efficiency versus contaminant particle diameter with the following chosen values: $r_\mathrm{f} = 10$ µm; $h = 2$ mm; $c = 0.3$; $C_\mathrm{m} = 20$ mass%; $v = 0.5$ m/s; kinematic viscosity of water 1 mPa·s and oil 20 mPa·s; amount of single-domain particles in agglomerate $n = 10$. Fiber filling by single-domain ferrite particles does not in fact change the filtration efficiency which is seen from the coinciding graphs for models I and II. Further magnetizing (model III) substantially increases filtration efficiency, which reaches a maximum in catching contaminant particles with $d_\mathrm{c} \geq 3$ µm in water filtration and $d_\mathrm{c} \geq 9$ µm in oil filtration.

Without magnetization, the efficiency of intercepting particles of the sizes mentioned does not exceed ∼25% (for water) and ∼60% (for oil). The use of an agglomerated magnetic filler (model IV) increases the filtration efficiency but insignificantly in contrast to unfilled fibers.

Table 8.1. Simulation of magnetic interaction between ferromagnetic contaminant particles and the filler in PFM

Model type	Force of magnetic attraction	Notations
II Single-domain non-magnetized filler	$$F(r) = \frac{4m^2 r_c^3}{l^7}$$	r, contaminant particle to fiber distance; m, magnetic moment of single-domain particle; l, average distance from ferrite to contaminant particle; r_c, contaminant particle radius
III Single-domain magnetized filler	$$F(r) = m_0 m \sum_{j=-N_j}^{N_j} \sum_{i=0}^{N_i} \sum_{\alpha=0}^{2\pi} \frac{r^2 + a^2 i + 2rai\sin\alpha + 4a^2 j^2}{(r^2 + a^2 i + 2rai\sin\alpha + a^2 j^2)^3}$$	m_c magnetic moment of contaminant particle; a, distance [a] between ferrite particles; $N_i = (r_f - r_p)/a$; r_p, ferrite particle radius; r_f, fiber radius; $N_j = \frac{l_f}{a}$, l_f, fiber half length; α varies with step $1/i$
IV Agglomerated magnetized filler	$$F(r) = nm_0 m \sum_{j=-N_j}^{N_j} \sum_{i=0}^{N_i} \sum_{\alpha=0}^{2\pi} \frac{r^2 + a_a^2 i + 2ra_a i\sin\alpha + 4a_a^2 j^2}{(r^2 + a_a^2 i + 2ra_a i\sin\alpha + a_a^2 j^2)^3}$$	a_a, distance [b] between agglomerates; n, number of single-domain particles in agglomerate; $N_i = \frac{r_f - r_a}{a_a}$; $R_a = r_p \sqrt[3]{\frac{n}{k}}$; k – ferrite particle packing factor in agglomerate; $N_j = \frac{l_f}{a_a}$

[a] Value a is found from the equation: $\dfrac{(r_f - r_p)(r_f - r_p + a)}{2a^3} = \dfrac{3}{8\pi} \dfrac{\rho_f}{\rho_p} C_m \dfrac{r_f^2}{r_p^3}$.

[b] Value a_a is found from the equation: $\dfrac{(r_f - r_a)(r_f - r_a + a_a)}{2a_a^3} = \dfrac{3}{8\pi} \dfrac{\rho_f}{\rho_p} C_m \dfrac{r_f^2}{r_p^3} \dfrac{1}{n}$, where ρ_f is the fiber density; ρ_p is the density of ferrite particles; and C_m is the ferrite content (mass%) in the PFM.

The main parameters characterizing magnetic filtering PFM are fiber diameter, FM density, ferrite concentration and particle size. Calculation dependencies for filtration efficiency versus the parameters named are shown in Figs. 8.3b–d and 8.4b–d for a contaminant particle size of about 5 μm. By reducing the fiber diameter in a type III FE from 40 to 6 μm for oil and to 20 μm for water, the filtration efficiency increases from 10 to about 100% (Figs. 8.3b and 8.4b). If the fibers do not contain magnetic filler, the dependence shifts to the side of lower fiber diameter values, i.e., to capture all particles being tested the fiber diameter should be below 3–4 μm which inevitably leads to increasing hydrodynamic resistance of the FE.

When the amount of ferrite filler is increased from 5 to 30%, the calculated filtration efficiency for oil in a type III FE rises from 30 to about 100% and for water from 60 to 100% (Figs. 8.3c and 8.4c). For complete purification of water from contaminants of more than 5 μm diameter, a ferrite concentration of about 20 mass% is sufficient. The filtration efficiency of type IV FE does

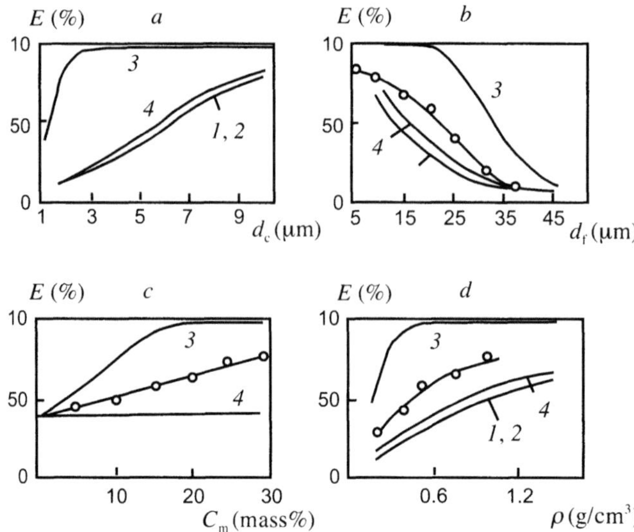

Fig. 8.3. Theoretical (without dots) and experimental (with dots) dependencies of the efficiency of water filtration through PE-based PFM on (**a**) contaminant particle size, (**b**) fiber diameter, (**c**) ferrite concentration, and (**d**) FM density [12]. Curves 1–4 correspond to models I–IV

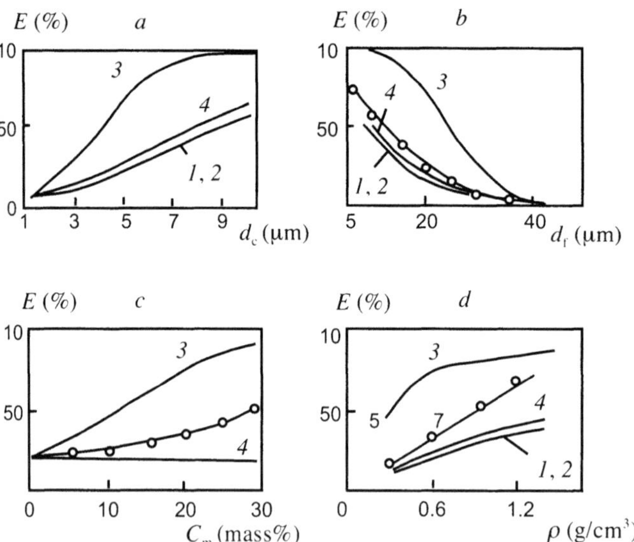

Fig. 8.4. Theoretical (without dots) and experimental (with dots) dependencies of the efficiency of mineral oil filtration through PA-based PFM on (**a**) contaminant particle size, (**b**) fiber diameter, (**c**) ferrite concentration, and (**d**) FM density [12]. Curves 1–4 correspond to models I–IV

not practically depend on the ferrite concentration in the fibers. The type III FE with a material density above 0.6 g/cm^3 completely cleans water and oil (Figs. 8.3d and 8.4d).

The reported analysis of calculation dependencies has proved these FE containing magnetized single-domain ferrite particles (model III) is preferable to the other types of FEs considered. In a real FM, however, the magnetic filler consists of a mixture of single-domain particles and agglomerates, so in calculations according to the formulas in Table 8.1 its worthwhile using an averaged n value.

8.3 Theory versus Experiment

The validity of using mathematical models for studying filtration efficiency has been checked by experiments. PE and PA-based filtering materials produced by the melt blowing technique underwent testing. Samples of 60 mm diameter were cut from a 2 mm thick FM sheet. The mean fiber diameter was 20 µm, the average particle number in agglomerates was four, and ferrite concentration in fibers was 20 mass%.

The filtration characteristics were estimated on a test bench by pumping contaminated liquid through FM samples. Liquid specimens before and after filtration were analyzed by an optical-magnetic detector [13]. Its operating principle is based on recording optical radiation which has passed through the specimen in an applied magnetic field (see details in Sect. 8.5). The total contamination of the liquid was recorded by relative optical density variation $D_1 = \ln(I_1/I_2)$, where I_1 and I_2 are transmittance of the pure liquid and that contaminated but cleaned by the filter, respectively. To determine the concentration of ferromagnetic particles, the relative optical density $D_2 = \ln(I_3/I_2)$ was found, where I_3 is the transmittance of liquid passed through the filter in an applied magnetic field. Then, using a calibrating plot, the concentration of ferromagnetic particles in the specimen was determined.

Water was used as a liquid carrier in testing a PE-based FM and mineral oil in testing a PA-based FM. The liquids were contaminated by carbonyl iron with a 0.5–12 µm dispersion (the mean particle size was 5 µm).

It is seen (Figs. 8.3b–d and 8.4b–d) that the experimental dependencies are located between calculation curves for model III (single-domain magnetized ferrite particles) and model IV (agglomerated magnetized ferrite particles).

8.4 Magnetization of PFM

The filtration efficiency for liquids containing ferromagnetic contaminants is conditioned by a set of factors involving ferrite concentration in PFM and

Table 8.2. Residual magnetic induction B_r of FM with different ferrite filler content (FM density, 0.3 g/cm^3)

Filler	B_r (mT) depending on filler content (mass%)			
	5	10	15	20
Barium ferrite	0.25	0.51	0.72	1.00
Strontium ferrite	0.32	0.65	1.03	1.33

its magnetization. Increasing ferrite concentration enables raising the values of the residual magnetic induction B_r of PFM (Table 8.2). The experimental values obtained are a bit less than the theoretical limits of magnetization. For example, the calculated maximum magnetic induction of PA-based PFM filled by 20 mass% of SF is 1.64 mT, and the experimentally achieved value is 1.33 mT (see Table 8.2).

As already stated, the ferrite concentration, FM magnetization, and particle size of a ferromagnetic contaminant influence the liquid filtration efficiency (Fig. 8.5). This is related to the increased probability of magnetically intercepting carbonyl iron particles with increasing residual magnetic induction values in response to growing ferrite concentration (curves *2, 4*). This effect is most evident in cleaning liquids with fine contaminant particles (curves *1, 2*).

It should be emphasized that when fine contaminants are to be captured, liquid filtration efficiency depends strongly on both FM density and the ferrite content in the FM (Fig. 8.6). Along with the intensification of traditional filtration mechanisms (inertial interception, gravitational precipitation,

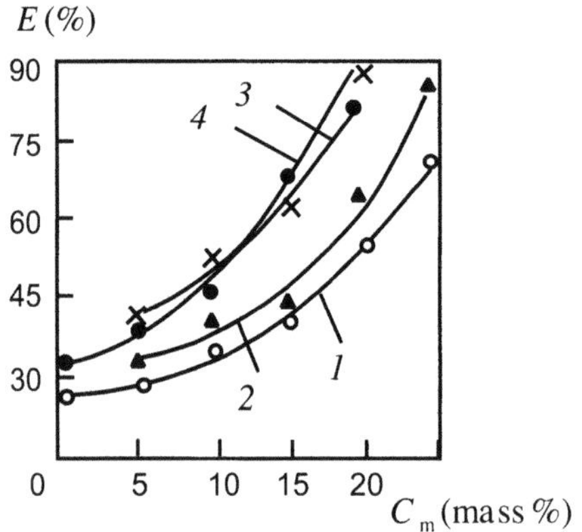

Fig. 8.5. The effect of the barium ferrite content in nonmagnetized (1, 3) and magnetized (2, 4) PA-based FM on the filtration efficiency of a machine oil containing carbonyl iron with particle size up to 12 μm (1, 2) and to 50 μm (3, 4). FM thickness, 2 mm

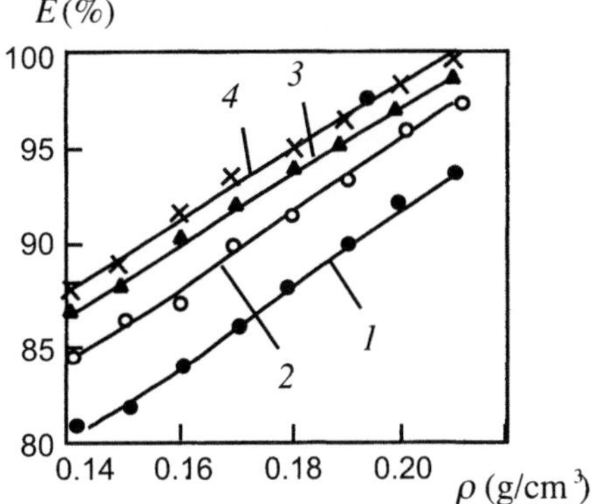

Fig. 8.6. The dependence of the filtration efficiency of machine oil contaminated by carbonyl iron with mean particle size 5.6 μm versus density of PA-based FM containing, respectively, 5 (1), 10 (2), 15 (3), and 20 (4) mass% of strontium ferrite. FM thickness, 2 mm

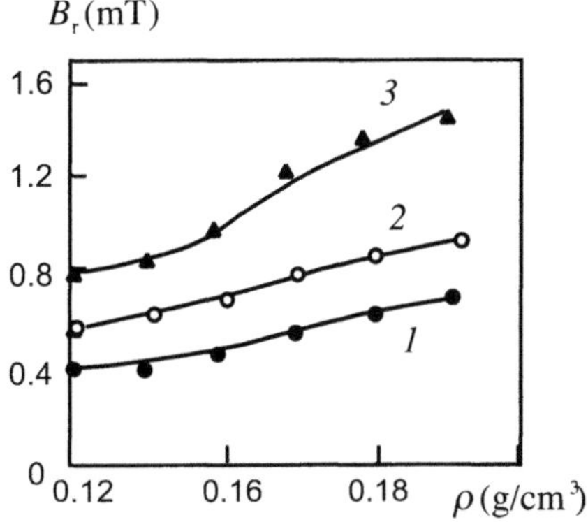

Fig. 8.7. Residual magnetic induction of PA-based FM depending on its density with various barium ferrite contents, mass%: 10 (1), 15 (2), 20 (3)

screening effect), increased FM density leads to shortening the distance between local sources of the magnetic field (magnetized ferrite particles), thus assisting in augmenting residual magnetic induction (Fig. 8.7) and increasing the magnetic capture of contaminant particles. The filtration fineness also depends on the FM density (Table 8.3). For comparison, analogous data for a pleated cardboard filter are given in the table.

One of the problems arising in processing polymer materials filled by magnetics consists of the following: ferrite particles tend to form aggregates retained by the forces of molecular and magnetic interaction. This hampers magnetic particle orientation in the polymer binder under the action of a texturing magnetic field, reduces the anisotropy of magnetic properties, and

Table 8.3. Various size particle content (in % of the total particle number) in oil samples upon filtration through FMs of different densities

Filtering material	Particle size, μm					
(density)	0–5	6–10	11–20	21–30	31–40	> 40
Magnetic PFM $(c = 0.18^a)$	100.00	–	–	–	–	–
Magnetic PFM $(c = 0.16^a)$	99.00	1.00	–	–	–	–
Pleated cardboard $(\rho = 0.54 \text{ g/cm}^3)$	96.53	2.60	0.43	0.22	0.22	–

$^a c$ is packing density, see explanations of formula (8.2).

limits the magnetic characteristics of a FM. By applying various methods of filler dispersion, a considerable increase in the amount of single-domain magnetic particles in a magnetic PFM can be attained [8].

In Fig. 8.8 the dependencies of magnetic induction on strontium ferrite (SF) content are presented in model samples based on paraffin (10 mm in diameter and 40 mm high cylinders) under various conditions of ferrite filler dispersion. Ultrasonic (US) ferrite dispersion in the binder melt partially breaks aggregates and makes filler particle distribution more even. As a result, further magnetizing of the sample yields higher B_r values in contrast to mechanical homogenizing of the blend (curves 1 and 2).

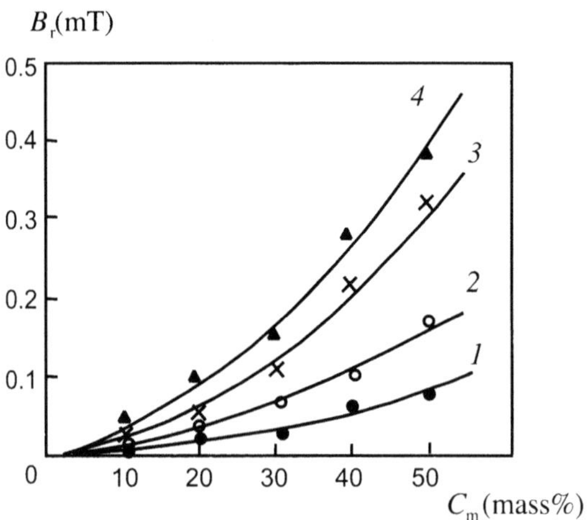

Fig. 8.8. The dependence of residual magnetic induction of a model paraffin-based sample versus SF concentration and dispersion procedure: 1, mechanical dispersion; 2, ultrasonic; 3, magnetic; 4, magnetic + ultrasonic

The intrinsic magnetic moment of a ferrite particle is rigidly bound to the axis of light magnetization; to change its direction by the action of a medium intensity magnetic field is possible only by turning the particle itself. This is achieved when the binder in a viscous-flow state is magnetized. The mold with paraffin + SF composition was placed inside a magnetizing coil ($H = 30$ kA/m) and heated up to the paraffin melting point. The composition was shaken up by periodical switching of short-term current in the coil, upon which the sample was cooled under a passing magnetizing current. A significant growth of B_r (curve *3*) was observed after this procedure. Even higher B_r values were obtained under a combined US and magnetic effect (curve *4*).

So, it is obvious that the additional use of US and magnetic treatment during magnetic PFM production helps to break magnetic filler aggregates and improve critical magnetic characteristics (and, consequently, filtration efficiency) of PFM owing to perfected magnetic anisotropy and particle orientation by the axes of light magnetization along the texturing magnetic field.

8.5 Magnetic Coagulation of Particles in PFM

A distinguishing feature of traditional commercial apparatuses for magnetic filtration of contaminated liquids is the availability of a strong inhomogeneous magnetic field. Interception of magnetic contaminants proceeds within the zone of the highest field intensity gradient. The major regularities of the processes named have been described elsewhere [9,14].

Magnetic PFM can be referred to as a new modification of magnetic separation systems. Pores and channels in PFM generate relatively weak (0.2–2 kA/m) but high-gradient magnetic fields (MF) active in filtration processes. When different media are passed through PFM, traditional mechanisms of depth filtration are realized, namely, inertial interception and screening, gravitational precipitation, diffusion, and impaction. However, investigations have proved that the decisive effect here is exerted by coagulation and interception of both magnetic and nonmagnetic particles under the magnetic field of the PFM itself.

Model dispersion systems approaching real contaminated media have been chosen to describe the phenomena of magnetic coagulation in industrial wastewater cleaning through PFM [15]. Suspensions of river sand atomized by a jet mill to $d = 1$–5 μm particle size and powdered nickel with $d = 5$–10 μm in dioctyl phthalate (DOP) and vacuum oil have been used. In addition, Charcoal suspensions ($d = 20$–50 μm) and DOP emulsions in distilled water stabilized by 0.1% addition of an anionic surfactant (sodium sulphoethoxylate), as well as hydrosols of chemically pure silicic acid at pH 9 (NaOH alkali agent), have also been studied. The initial concentrations of the dispersed phase in sand suspensions were 0.075 and 0.15, and in the

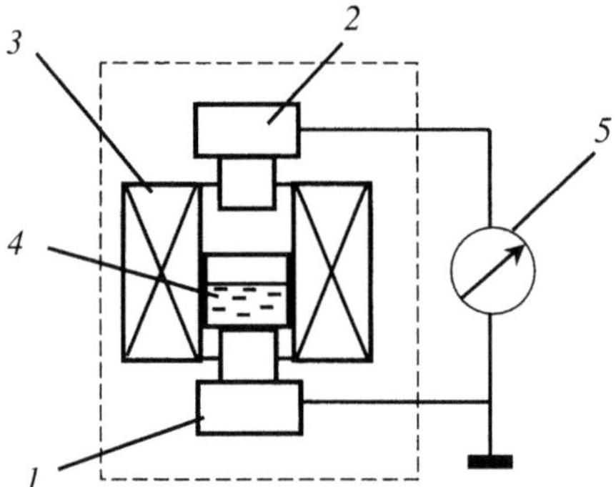

Fig. 8.9. A diagram of a magneto-optical analyzer: 1, radiator (IR diode, $\lambda = 950$ nm); 2, detector with photoreceiver and preamplifier; 3, electromagnetic coil; 4, vessel with liquid sample; 5, voltmeter

remainder, 0.075 mass%. Before dispersion, sand particles were electrified in a negative corona discharge field. The efficient surface charge density ESCD (σ_{ef}) of the particle layer was estimated by a contactless induction method [16].

The processes of separation and coagulation in dispersed systems were examined by a magneto-optical analyzer (Fig. 8.9) whose operating principle is based on recording variations of light signals passing through the dispersed system in an applied spatially inhomogeneous MF. Also considered was the field intensity gradient $grad_r H$ in the radial direction of a vessel containing a liquid sample inside an electromagnetic coil.

The optical density of suspensions and emulsions varying in time was calculated by the formula

$$D(t) = -\ln[U(t)/U_0] \,, \tag{8.3}$$

where $U(t)$ and U_0 are measured by detector voltage values corresponding to optical radiation intensity through a dispersed system and a pure dispersed medium.

It was established that Ni and sand suspensions in oils (DOP and vacuum oil) have a relatively high aggregation and sedimentation stability in the absence of an external MF. This is, in particular, expressed in a rather low variation of suspension optical density with time (Fig. 8.10, curves 1, 1'). The important factors for the stability of these lyophobic dispersed systems are, evidently, low dispersed phase concentration and high dispersive medium viscosity. Along with this, electrified sand suspensions in oil are stabilized through electrostatic repulsion of solid particles whose surface layer bears a negative electrical charge. The charge in media unsusceptible to ionic or electron conductivity is preserved for a relatively long time. Its to be considered that because of intrinsic dipolar moment in DOP molecules and their

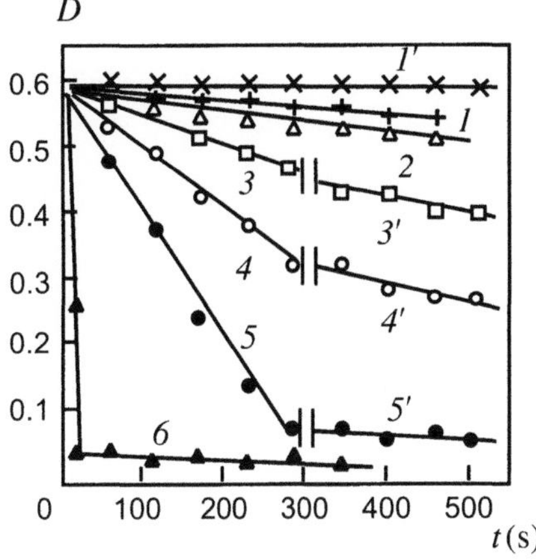

Fig. 8.10. Optical density D variation of sand (1–5) and nickel (6) suspensions in DOP without (*1, 1'*) and with an inhomogeneous MF (*2–6*). Field intensity in the coil center $H_0 = 24$ kA/m. The ESCD of sand particles (nC/cm^2): 0 (1, 2); 1 (3); 2 (4); 3 (1', 5). Curves 3', 4', 5' upon turning off the MF

ability to polarize, oriented adsorption of DOP dipoles on the particle surface and formation of a double electrical layer (DEL) occurs.

As soon as the electromagnet is actuated, the system stability is violated. The optical density of Ni suspensions drops within a short time to zero (Fig. 8.10, curve *6*). Magnetic particles start to migrate intensely to the vessel side walls, where they coagulate and precipitate upon field removal (Fig. 8.11a). Sand particles travel, though slower by far, in the $grad_r H$ direction, too, and the circular zone of sedimentation remains rather wide (Fig. 8.11b). The linear variation of D with time, i.e., the independence of the process rate on particle concentration, shows that coagulation goes on beyond the limits of the detection region (that is, the sample volume affected by the analyzer light flux). Angle indexes k of linear dependencies $D(t)$ characterize the rate of particle number reduction (per unit detected volume) induced by their magnetic drift.

The drift, it has been noticed, accelerates when the polarizing charge of the particles and the MF intensity increase (Fig. 8.12). Migration is impeded by the high viscosity of the dispersed medium. For example, k values for charged sand in vacuum oil are, on average, $3.3 \pm 0.3 \cdot 10^{-4}$ s^{-1}.

Based on the results obtained, *the mechanism of magnetic coagulation processes* can be presented as follows. A directed motion of particles occurs in the liquid due to the action of magnetic force F_m compared to which other forces (inertial, gravitation, etc.) can be neglected (see Sect. 6.2). When a charged particle moves across MF force lines, it experiences Lorentz force F_L, which distorts the particle trajectory. Calculations have, nevertheless, proved that $F_L \ll F_m$, so both particle travel velocity and behavior are

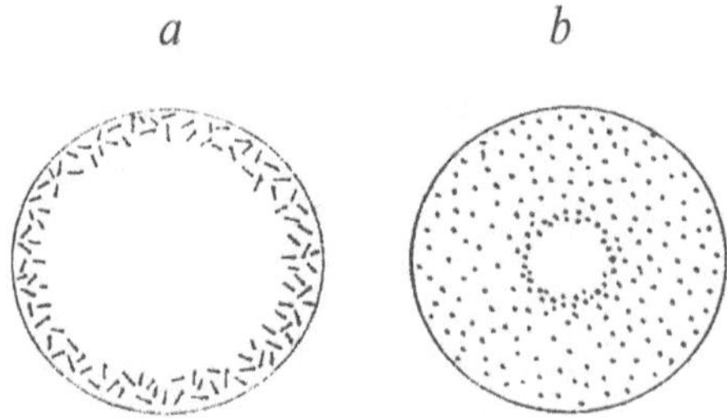

Fig. 8.11. Distribution of coagulated sediments for Ni (**a**) and sand (**b**) on the vessel bottom upon the inhomogeneous MF action on dispersed systems (dispersion medium, DOP)

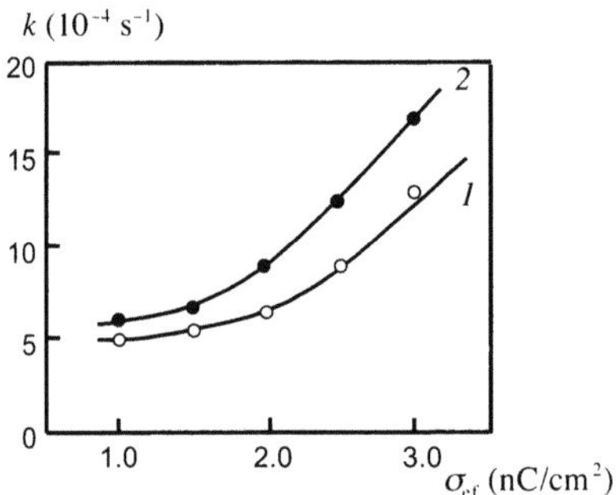

Fig. 8.12. Effect of sand particle ESCD and MF intensity in the coil on the constant of the reduction rate of particle concentration within the optical detection zone. H_0 (kA/m): 12 (1) and 24 (2)

defined by relation F_{m}/γ, where γ is the Stokes coefficient that depends on liquid viscosity, particle size, and shape.

Ni is quickly separated from suspensions in a gradient MF owing to high magnetic sensitivity of this substance [17]. In contrast to Ni, sand particles acquire just a slight magnetic moment because of admixtures of antiferromagnetic α-Fe$_2$O$_3$ (up to 7%) of a geologic nature and ferromagnetic Fe$_3$O$_4$, probably introduced into the sand during milling. Magnetite is, in fact, an inevitable product of the corrosion steel part in technological equipment and a most common contaminant encountered in technical media [9].

Weakly magnetic particles migrate in the $grad_{\mathrm{r}}H$ direction very slowly (Fig. 8.10). If preliminarily electrified, their drift velocity grows significantly. Supposedly, electrical polarization of the sand reconstructs the magnetic structure of admixtures that bring about magnetization $I = \beta E$ (β is coefficient) whose value depends on the affecting electrical field intensity. So, the magnetic moment of particles increases, and their drift velocity in a gradient MF also increases. A magnetic electrical effect is also observed in antiferromagnetics whose magnetic structure is similar to α-Fe_2O_3 and in magnetite (Fe_3O_4) [17,18].

Magnetic treatment initiated coagulation with further phase separation in the microheterogenous systems under investigation. The mechanisms of coagulation processes in high and low-magnetic dispersed phases are different, however.

In the former case, the high concentration of magnetic particles (Ni) is reached within the zone of maximum intensity of the MF. The interaction energy (W) of magnetic dipoles (μ) located close (l) to each other increases significantly because $W = \mu^2/l^3$. Thus, chains of magnetic particles are formed that are oriented along the field intensity lines [13]. Chain aggregates start to sediment upon MF removal.

Individual slightly magnetic sand particles can also experience magnetic attraction when traveling in the medium and approaching each other under the $grad_r H$ effect. But a more important coagulation factor in such systems is interaction between electrical dipoles that results from DEL deformation of moving particles. The effect of long-range forces of attraction between induced dipoles causes the formation of particle aggregates, probably, chain aggregates. Loss of aggregate stability makes a suspension sediment, which is observed, even in the zone neighboring the detection region. This type of coagulation is irreversible: upon MF removal, the optical density of suspensions continues to be impaired (Fig. 8.10).

Now, let us discuss the phenomena arising in aqueous systems with *a diamagnetic* (either solid or liquid) dispersed phase under inhomogeneous MF action.

The dispersion of hydrophobic solid particles (coal) or oil drops (DOP) in aqueous solutions of sodium sulphoethoxylate is accompanied by anion adsorption of this surfactant on dispersed particle surfaces. The particles gain a negative charge and attract an equivalent number of Na^+ cations in the solution. Thus, an adsorptive DEL is formed that displays electrostatic and mechanical-structural barrier properties that prevent coagulation.

Electrostatic stabilization of silicic acid hydrosols is chiefly a result of the dissociation of surface silanol groups upon NaOH addition to the water base (resulting pH $= 9$) [19]. In this case, DEL formation can be attributed to a layer of SiO_3^{2-} and SiO_3H^- ions on the particle surface, as well as to diffusive concentration around it of Na^+ counterions in the liquid.

When a MF is absent, infringement of the dispersed phase distribution within the dispersion medium due to insufficient kinetic stability of the systems under study is expressed by the variation of their optical densities (Fig. 8.13a–c, curves *1*). Nevertheless, any coagulation phenomenon in the samples was not recorded during the experiment (to 10 min).

When the coil was actuated, diamagnetic particles of coal, silicic acid, or DOP droplets that have negative magnetic receptivity start to move under F_m action opposite to the $grad_r H$ direction. Their concentration in the

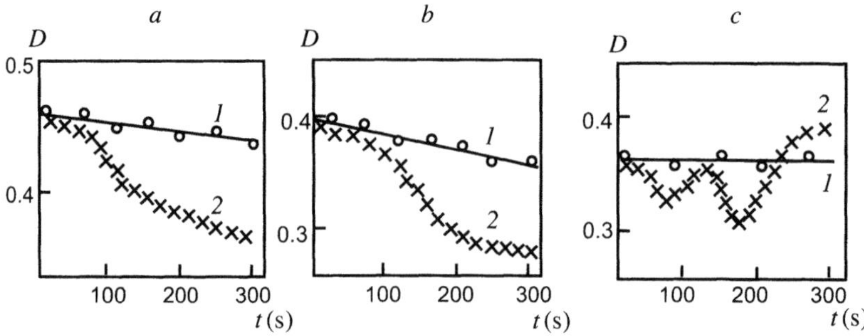

Fig. 8.13. Variation in the kinetics of optical density D of silicic acid (**a**) and coal (**b**) suspensions, as well as DOP emulsion (**c**) in distilled water with additions of stabilizers: NaOH (**a**) and 0.1% sodium sulphoethoxylate (**b, c**). 1, MF is absent; 2, inhomogeneous MF; $H_0 = 24$ kA/m

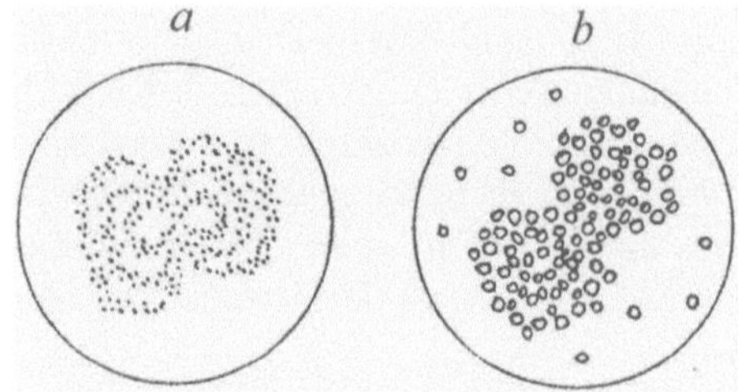

Fig. 8.14. Distribution of silicic acid coagulation sediments (**a**) and floating up drops of DOP (**b**) upon the inhomogeneous MF effect on dispersed systems (the dispersing medium is water)

central portion of the vessel increases. Nevertheless, the optical density of suspensions decreases (Fig. 8.13a,b) because solid particles form aggregates and precipitate (Fig. 8.14). The coagulation mechanism is related to DEL deformation in response to particle drift and the appearance of nonequilibrium electrical surface forces which bring about oriented aggregation. The behavior of curves $D(t)$ in Fig. 8.13a,b confirms that the coagulation rate depends on the particle concentration, and the formation of particle aggregates terminates within the time of detection.

A more intricate picture (Figs. 8.13c and 8.14b) is observed in DOP-water emulsions. The optical density of the sample placed in a homogeneous MF periodically drops and rises. Presumably, when oil drops migrate in an aqueous medium under the F_m effect, the DEL of the drops becomes polarized. The dipole interaction of DELs leads to the coalescence of drops. In addition, DEL polarization generates mechanical forces which stretch the drops [20]. The drops take the shape of stretched spheroids and break under some critical stress. This is characteristic of coarse drops whose surface tension forces are insufficient to compensate for stress in tension. Thus, the coalescence of fine drops is followed by the breakage of coarser ones. As can be seen, the nonequilibrium processes described violate the kinetic stability of emulsions and lead to separation of coagulation products of the organic phase.

Aside from the aforementioned, coagulation processes in the aqueous suspensions and emulsions under study are supposed to be activated by electromagnetic microhydrodynamic phenomena in the boundary layers of charged particles migrating in a MF [21]. The reason for such phenomena is in the oppositely directed ion charge sign and value of Lorentz forces which affect anions and cations situated, respectively, in the dense and diffusive parts of the DEL. That means that cations and anions gain a tendency to come apart which is hindered by their coulombic attraction. As a result, local potential differentials and magnetic hydrodynamic strains occur at the interface in the diffusive parts of the DEL. These phenomena might bring about deformation or even violation of particle solvate layers and surfactant adsorption films which promote coagulation and particle coalescence.

As can be seen, coagulation of suspensions is irreversible. Coarse drops formed by coalescence are deformed and start to break upon motion. The loss of aggregate stability under an inhomogeneous MF leads in all cases to phase separation in dispersed systems.

The processes considered evidently take place in filtration through magnetic PFM. A local high-gradient MF are generated in PFM, and their configuration depends on the material structure and magnetizing conditions. Magnetic coagulation phenomena increase the efficiency of other mechanisms of depth filtration and improve contaminant capture from liquids flowing through PFM.

8.6 Magnetic Capillary Phenomena

One of the factors of governing PFM filtration efficiency is fiber wetting by the filtered liquid which is interrelated with capillary phenomena in the filtering material bulk.

The magnetic field effect on liquid wetting and spreading on solid surfaces was recorded in the 1960s [22]. Interpretation of those effects has evoked a great deal of contradiction often in disagreement with established notions in physics. Nevertheless, investigations into the magnetic field effect on liquids in engineering systems and biological objects continued and their results were commercialized afterward [23–25].

Interest in these phenomena rests on their novelty and also on the ecological safety of a magnetic field as a technical factor. Its use has arisen because of the possibility of reducing the amounts of chemical reagents employed in industrial processes, of simplifying machine designs, and of prolonging filter service life.

The effect of an inhomogeneous MF on the capillary permeation of aqueous systems in PFM pores has been studied on the experimental system

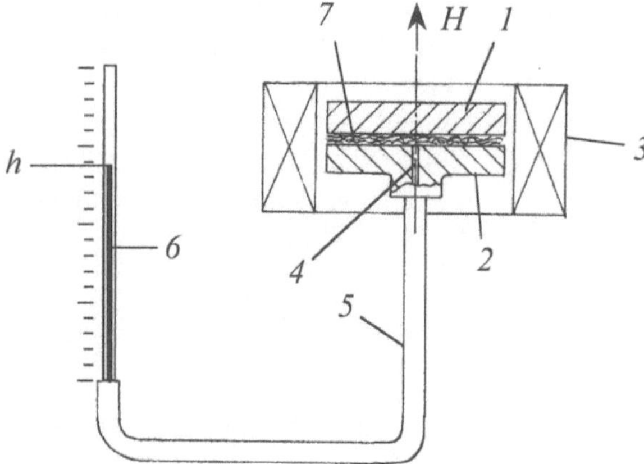

Fig. 8.15. A diagram of an experimental system to investigate capillary permeation of water in PFM. See explanation in the text

shown in Fig. 8.15 [26]. A pair of disks (*1*, *2*) 42 mm in diameter made of a nonmagnetic material was placed inside a circular solenoid *3*. The axes of the disks and that of the solenoid are aligned, so that the force lines of the solenoids MF can be taken as if it is directed along the normal to the disk surfaces. In the lower disk *2*, there is a channel *4* connected by tube *5* with a measuring glass capillary *6*. Tubes *5* and *6* are filled with water upon upper disk removal. Sample *7* (250 μm thick) from a LDPE-based magnetic PFM filled by 15 mass% of ferrite is installed between the disks. The PFM fiber diameter was 15–20 μm, and the porosity was 40%. Under capillary pressure, water from channel *4* fills the PFM pores. The process kinetics is evaluated by the liquid column height *h* drop in measuring capillary *6*.

In Fig. 8.16, the following cases of a MF effect on water flow in the capillary system under study are depicted.

A. The sample is not magnetized; the solenoid field is either switched off (curve *1*) or switched on (curve *2*). Ferrite particles in a nonmagnetized sample show a natural spontaneous magnetization. The magnetic moments of particles orient in the external MF direction after switching the solenoid on. The total MF of the solenoid and the sample accelerates the capillary motion of water as reflected in the location of curve *2* below curve *1*.

B. The sample is magnetized in a direction normal to the surface, and the solenoid field is removed. The kiinetic curves *3* of capillary flow coincide at any position of the magnetic poles of the sample relative to the disks. The requisite condition for the magnetic treatment of liquids is said to be crossing of MF force lines by the liquid flow [24]. So long as the force lines of the sample MF are always normal to the disk surface, capillary flow between disks crosses them similarly at any location of the poles. Hence, change of the force lines to the opposite direction does not influence the efficiency of the magnetic treatment of water.

C. The sample is magnetized, solenoid switched on. The behavior of kinetic curves (*4, 5*) of water flow is conditioned by the intensity of the total MF of the solenoid and sample. The effect of the magnetic treatment of water is stronger in similar field directions of solenoid and sample (*5*) and is weaker in the opposite (*4*).

The experiments described with running water containing considerable amounts of ferromagnetic admixtures were repeated with distilled and double distilled water at sample magnetization $B_r = 0.2$ mT. The difference in the velocity of capillary spreading in a magnetic field and without, it has been found, increases with an increasing degree of purification of water.

The results obtained can be explained by varying properties of water flowing in the clearance between disks through PFM pores and crossing MF force lines. This is, probably, one of the manifestations of the magnetic treatment of water [24].

The liquid flow rate in a horizontal capillary is known to be directly proportional to the cosine of the wetting angle, the liquid surface tension, and inversely proportional to the liquid viscosity [27]. Let us analyze how these water parameters vary under the influence of a MF.

It was definitely determined that the variation of the MF intensity from 0 to 100 kA/m did not change the water surface tension, e.g., double distilled [24], whereas the magnetization of water worsened solid surface wetting [24,

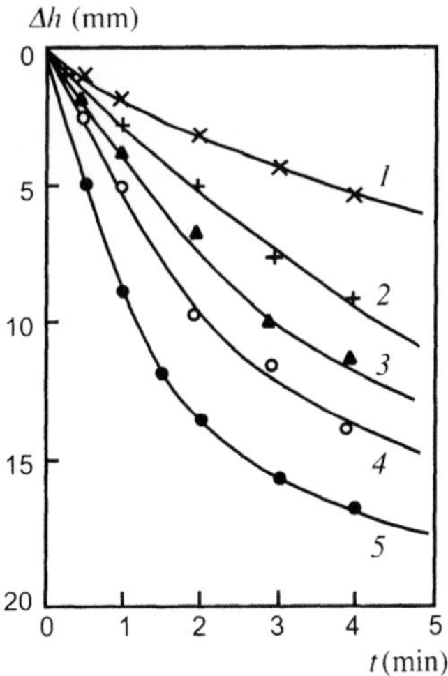

Fig. 8.16. Kinetics of water level variation in a measuring capillary when a sample of PFM is placed between brass disks having the following magnetic induction B_r (mT): 0 (1, 2) and 0.2 (3–5); *1, 5* – the solenoid is switched off ($H = 0$); *2, 3, 4,* the solenoid magnetic field is switched on ($H = 0.7$ kA/m)

28]. Contradictory information is available in the scientific literature on liquid viscosity variation in response to magnetic treatment [24, 25, 29]. It has been experimentally established that the capillary flow of water is accelerated by a MF. This proves that when the edge-wetting angle is really enlarging, the water viscosity will reduce. Moreover, the effect of viscosity reduction on the flow rate is stronger than that of wetting impairment. To interpret the results, let us examine the effect of water purification on these two parameters.

Wetting occurs when water touches the surface of a PFM sample and depends rather on the molecular properties of water than on those of the admixture ions. The latter concentration in the surface layer of running water is insignificant. Therefore, the variation in the edge-wetting angle from magnetic treatment does not, in fact, depends on the degree of water cleaning. With a certain amount of approximation, wetting parameters could be taken as constants in experimental conditions.

It will be convenient to consider the effect of contaminants on water viscosity from the viewpoint of representations on clathrate hydrates [30]. When coupled through hydrogen bonds with one another, water molecules form a three-dimensional skeleton in whose voids admixture molecules and ions are found. Electronic interaction between them and water molecules (hydration effect) is a more strong structure-forming factor than the association of water molecules due to hydrogen bonds. This is the reason that viscosity reduces within the "running water-distilled-double distilled" series. Naturally, all conditions being equal, the velocity of capillary penetration of liquids increases in this series. The effect of magnetic treatment which breaks clathrate hydrates becomes stronger with increasing cleaning of water. The viscosity of running water is changed less then that of double distilled after magnetic treatment; influence on capillary penetration is, respectively, increased within the "running water-distilled-double distilled" series.

The supposition is just one in a series of hypotheses about the MF effect on water systems which can be categorized in three groups:

1. A MF affects water molecules and permolecular structures.
2. A MF influences ions and other contaminants in water.
3. New compounds are formed in water by magnetic treatment.

In the hypotheses of *the first group* [31], the MF is presumed to generate Larmor precession of electron orbits and nuclei along with polarization of electron clouds in water molecules thanks to which the molecules gain an induced moment antiparallel to the external field. As a result, partial bending and disruption of hydrogen bonds takes place. The number of molecules leaving quasi-crystalline lattice sites and filling their cavities grows.

Yashkevich's model [32] subdivides aqueous molecules into two types. Molecules constituting the base of the liquid water skeleton (i.e., molecules with four hydrogen bonds) form a subsystem "quasi-skeleton", and the remaining molecules carrying three and lower hydrogen bonds are related to the subsystem "structural defects". The molecules can transfer from one subsystem into another. A MF might either assist in formation of hydrogen bonds ("birth" of a quasi-skeleton) or, vice versa, stimulate the "destruction" of a quasi-skeleton when hydrogen bonds start to rupture thus either elevating or lowering water activity.

The second group of hypotheses, where the authors try to isolate certain admixtures and make them responsible for the phenomenon as a whole, are more numerous.

A supposition has been put forward in [33, 34] that the magnetic treatment' of aqueous solutions becomes efficient under two conditions: solution saturation by a scale-former and the presence of ferromagnetic particles in it. Ferromagnetic particles are magnetized in a MF and start to aggregate. The aggregates sorb ions and molecules of the scale-former on their developed surfaces. These were conclusions based on experiments with water containing ferromagnetic particles.

Other investigators assigned the leading role in interactions with a MF to ions. Yu.V. and I.V. Myagkov [35] supposed that ions form quasi-crystalline ordered sublattices from cations and anions with more than 10^6 number of ions. The magnetic field brings about a head-on drift of cation and anion sublattices.

The second group also includes hypotheses asserting that magnetic treatment is reduced to degassing of the liquid [21, 24]. Lorentz force affecting charged gas bubbles is neglected in calculations in contrast to the Stokes resistance. Nevertheless, flow pressure changes jumpwise under gradient hydrodynamic flow (i.e., flow with a variable cross section), and especially in periodical variation of a capillary section. Gas microbubbles begin to pulse intermittently constricting and extending. Coulombic and van der Waals mutual attraction forces act between the charged pulsing bubbles. When a MF is actuated, a strong effect is exerted by Lorentz forces which shift charge carriers of different sign surrounding each of approaching bubbles to opposite sides. Elementary charges (or ions) change position and fall behind the moving bubble wall at the moment of contraction. When the charge is removed from the bubbles, the coulombic force of repulsion is eliminated, and the bubbles stick together from the force of attraction.

The third group of hypotheses attributes all phenomena occurring at magnetic treatment of aqueous solutions to formation of new compounds not found previously in water. According to [36], hydrogen peroxide, ozone, nitrogen oxides, and other active substances are formed in water by magnetic treatment. It was shown that the amount of hydrogen peroxide increases in water after magnetic treatment by $1.5 \cdot 10^{-5}$ mol/L. But other authors could not confirm this result [24].

The previously cited hypotheses do not contradict, in principle, experimental results cited earlier. Moreover, each hypothesis can be successfully used to explain these results and prove the reliability of the proposed model of water filtration through magnetic PFM.

8.7 Serviceability of Magnetic PFM-Based Filters

Filter performance and service life depend strongly on the pressure difference in the filtered medium passing across a FE. The filter was tested by passing liquid through it once [5, 6] (Fig. 8.17). PFM filling with ferrite and magnetizing brings about just a negligible pressure rise of the liquid before the filter (pressure after the filter is, as a rule, constant). This regularity is weaker when coarse particles (below 50 µm) of carbonyl iron are filtered. Finer particles probably penetrate deeper into the FM and block its pores.

When liquids containing nonmagnetic contaminants are filtered through ferrite-filled PFM, filtration rating values are 7–15% higher compared to unfilled PFM. As noted previously, nonmagnetic contaminants precipitate in FM mainly by the mechanisms of inertial interception, gravitational precipitation, and van der Waals interaction. Magnetization initiates additional particle coagulation and screening, although the pressure difference increases negligibly (Table 8.4).

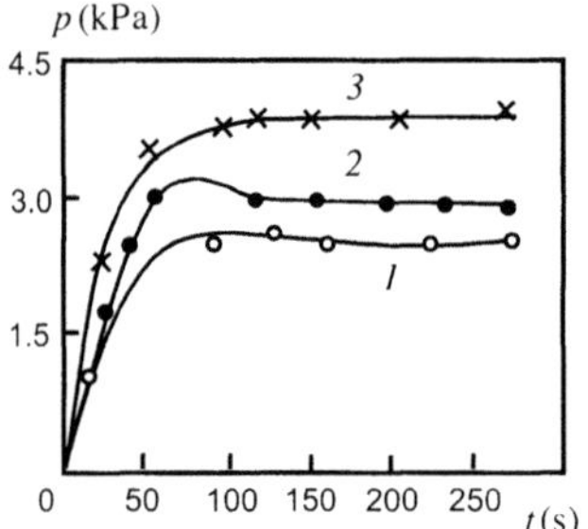

Fig. 8.17. Variation of oil pressure with time before a Pa-based PFM without filler (*1*) and filled by 15% SF (*2, 3*): *2*, non-magnetized sample; *3*, magnetized up to $B_{\mathrm{r}} = 1.2$ mT. Carbonyl iron particle size up to 12 µm; filtration rate, 1 L/min

Table 8.4. Pressure difference (Δp) in filtration of mineral oil containing nonmagnetic particles ($d_{\mathrm{c}} < 50$ µm)

	Δp (kPa) across filtering element	
Particle material	Nonmagnetized	Magnetized
Quartz sand	6.0	7.0
Aluminosilicate enamel	6.5	7.5
Porcelain	7.0	8.0

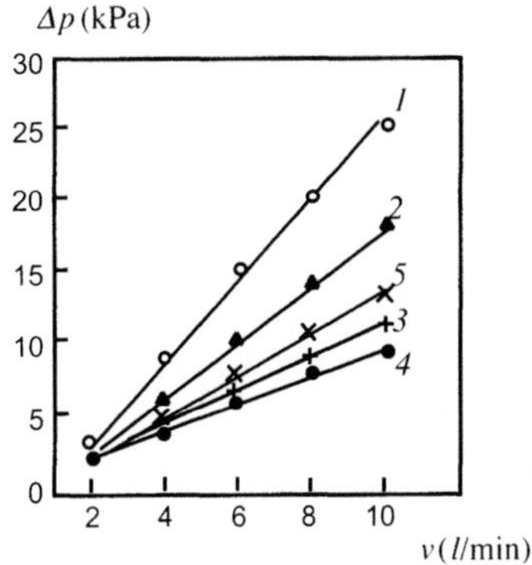

Fig. 8.18. The effect of oil filtration rate on pressure difference across a FE made of PA-based PFM (barium ferrite content 15 mass %) with various packing densities (1–4) and across a paper FE with the same dimensions (5). Packing density: 0.139 (1); 0.125 (2); 0.111 (3); 0.097 (4)

Table 8.5. Fuel consumption and oil pressure in an internal combustion engine (ICE) in filter bench tests

Regime of ICE operation		Fuel consumption (L/h)			Oil pressure (kPa)		
Rotation n	Torque	For PFM with packing density		For paper filter	For PFM with packing density		For paper filter
(rpm)	(N·m)	0.111	0.125		0.111	0.125	
1650	18	2.05	2.06	2.06	180	170	170
2350	22	3.33	3.35	3.35	270	260	262
3800	33	6.6	6.64	6.63	340	325	328

Combined investigations in the serviceability of oil filters based on magnetic PFM have been carried out at the Korea Institute of Science and Technology (KIST) [12]. The pressure difference was measured for a FE made of PA-based PFM with varied packing density and fixed barium ferrite content (15 mass%). The pressure difference, it was noticed, grew with intensified filtration rate and increased material density (Fig. 8.18). It is obvious that the pressure difference across a conventional paper filter is approximately the same as that across a PFM-based FE that has a packing density of 0.111 (curves *3* and *5*).

The prolonged serviceability of filters was examined on a test bench for studying internal combustion engines (Benz M102E). Motor oil 10W/30 was filtered for 120 h which corresponds to 10,000 km of automobile mileage. It was established that a PFM-based filter with 0.125 and 0.111 packing density intercepts particles above, respectively, 5 and 20 μm, whereas a paper filter removes particles that exceed 40 µm from oil. It was also noticed that the metal wear debris content (Fe, Cr, Cu, and Pb) in waste oil was 1.2–1.8 times less in a fibrous filter application compared to that using paper filters. At the same time, substitution of paper oil filters for fibrous filters does not influence fuel consumption or oil pressure in the engine (Table 8.5).

The results of bench tests confirmed earlier conclusions that a difference exists between the filtration mechanisms of filters made of paper and PFM. The efficiency of paper filters is fully determined by material porosity and depends on the ratio of pore size to entrapped debris dimensions. In magnetic fibrous filters, the capture of fine metal particles is effected by the internal magnetic field of a FM.

9. Adsorptive and Microbicidal PFM

Adsorptive PFM are designed for combined deep filtration of industrial sewage where fine suspensions of solid particles, emulsified petroleum products, dissolved heavy metal salts, organic toxicants, and detergents are present simultaneously in significant variations in pH and waste composition.

Such FM contain adsorptive substances immobilized in a polymer fibrous matrix, including

- highly-porous carbon and other inorganic adsorbents (natural and synthetic active coal, carbon fibers, zeolites, aerosil, etc.);
- ion-exchange polymer fibers (based on modified PAN, PVA, PA) and some other types of ionites;
- complexing agents (crown esters, nitrogen-containing heterocyclic compounds, ferrocyanides, and so on).

Porous adsorbents might be impregnated with deodorizing and aromatic compositions. In this case, deodorizing PFM are used in cleaning gas emissions.

9.1 PFM Modified by Porous Adsorbents

A distinguishing feature of these PFM is adhesive fixing of adsorbent particles on fibers (Fig. 9.1a). It is not expedient to process adsorbents simultaneously with polymer extrusion because porous particles become encapsulated by the binder and thus lose their adsorptive ability (Fig. 9.1b). If adsorbent particles are injected into the gas-polymer flow during PFM formation, they collide with and adhere to fibers which are in the viscous-flow state and firmly fix on the fiber surface upon solidification [1, 2].

Upon computer processing of electron microscopic images (Fig. 9.1a), it was found that the particles of 13.8 mass% activated charcoal (AC) content occupy about 37.8% of the fiber surface area (the mean particle size and fiber diameter d_f are commensurable). Evidently, to cover 100% of the fiber surface with AC particles, the maximum amount introduced into the PFM should not exceed 35–40 mass%. In reality, it is probable that introducing less than 20–25% of modifier will not affect its adhesion to fibers.

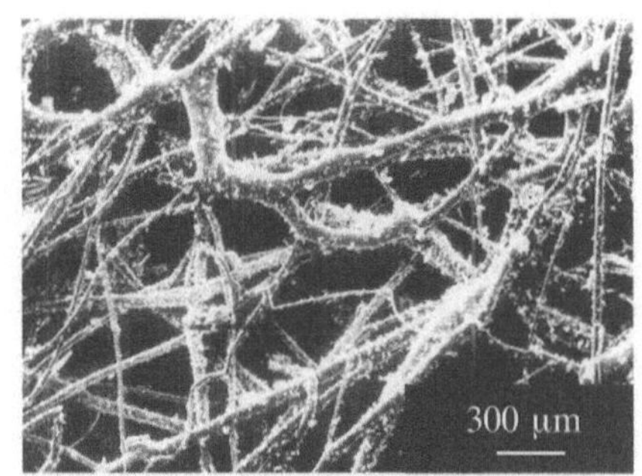

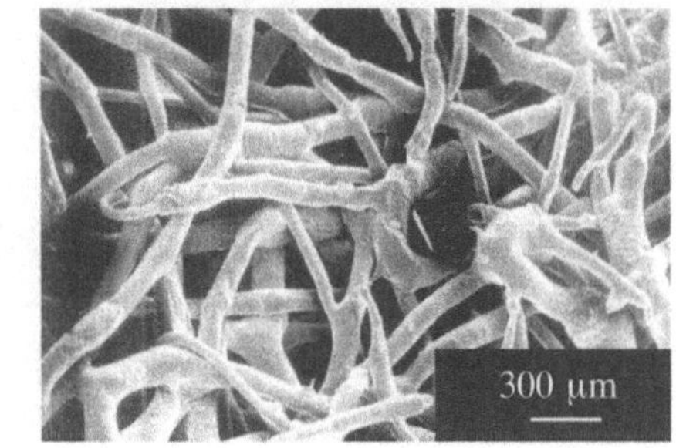

Fig. 9.1. Electron microphotographs of LDPE-based PFM modified by AC (13.8 mass%): (**a**) charcoal particles injected into the gas-polymer stream; (**b**) charcoal particles extruded simultaneously with the polymer melt

Filtration of various media through PFM modified by adsorbents is accompanied by the capture of both solid and liquid dispersed contaminants as well as by adsorptive concentration of dissolved organic and inorganic admixtures. When a PFM follows the mechanism of physical sorption, the parameters of its porous structure of importance are, the volumes of sorption (V_s), the micro- (V_{mic}) and mesopores (V_{mes}), the mean efficient pore diameter, and the specific surface of sorption (S_{sp}).

The sorptive characteristics of AC and PFM modified by it have been calculated using sorption isotherms of benzene vapors (Fig. 9.2). It was established that micro- and mesopores of AC and mesopores formed in the gaps (openings) between fibers and charcoal particles at contact sites in composite PFM are adsorptive. The characteristic size of the mesopores formed is 250-350 Å which is in conformity with a definitely expressed peak on differential curves of mesopore distribution by size. Maximum corresponding to this pore size is absent on an analogous curve for AC.

Analysis of the data presented in Table 9.1 proved that the method of introducing porous adsorbents into melt-blown PFM excludes encapsulation by the binder and achieves materials with highly adsorptive characteristics. PFM containing 11–14 mass% of AC possesses a developed surface S_{sp} that is three orders larger than that of an unmodified surface, and its volume of

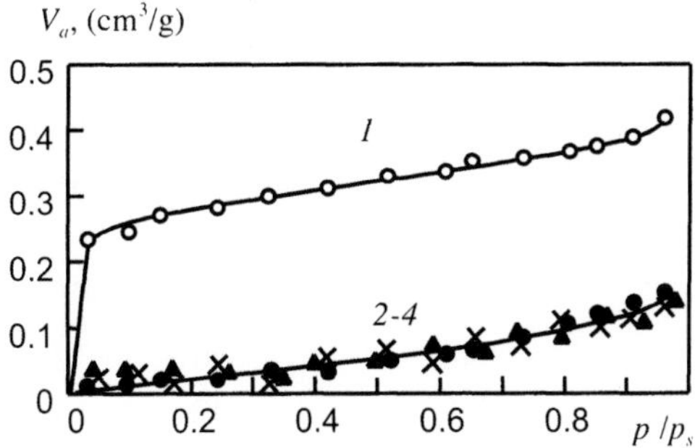

Fig. 9.2. Isotherms of benzene vapor sorption at $T = 20°C$ on AC (1) and PFM with AC content (mass%): 13.8 (2), 12.6 (3), and 10.8 (4). V_a (cm^3/g) is the adsorbate volume in a liquid state; p/p_s is the relative pressure of the adsorbate vapor

Table 9.1. Sorption characteristics of AC and LDPE-based PFM modified by AC

Sample	Sorption capacity SC$_{MB}$ (mg/g)	Specific surface S_{sp} (m^2/g) by BET	Pore volume (cm^3/g)		
			V_s	V_{mic}	V_{mes}
AC	167	400.4	0.415	0.309	0.156
PFM with AC content (mass%):					
13.8	–	63.3	0.139	0.048	0.132
12.6	10.7	57.5	0.133	0.043	0.129
10.7	6.5	64.4	0.154	0.043	0.155
0	0.1	0.02–0.1[a]	–	–	–

[a] by sorption isotherm of nitrogen at different PFM fiber diameters.

sorptive micro- and mesopores is rather appreciable. The PFM adsorption index of methylene blue (SC$_{MB}$) from aqueous solutions increases by two orders upon modification with charcoal.

9.2 PFM as Adsorbents of Oil Products

In some cases the efficiency of adsorptive filtration is conditioned by the nature and structural parameters of the fibrous matrix itself (P, d_f, S_{sp}). Note that melt-blown materials consisting of thin lipophilic fibers (PE, PP) are perfect petroleum absorbers [3]. Their petroleum-retaining capability in static conditions reaches 10g/g and far surpasses, in this respect, composite materials intended to pick up oil products (Table 9.2). A promising sphere of PFM

Table 9.2. Static adsorptive capacity for oil products (SC$_{OP}$) depends on PFM composition and structural parameters

Material and its composition (mass%)	Structural parameters		SC$_{OP}$ (g/g)
	Mean fiber or adsorbent particle diameter (μm)	ρ(kg/m^3)	
Melt-blown PFM based on			
PE fibers, 100	30	300	8
PP fibers, 100	20	140	4
PP fibers, 100	15	100	10.4
PP fibers twisted into			
braids and balls [5]	5–20	10–200	16–53
Petroleum sorbents produced at 3M Co. on PP fiber base	–	–	8–15
Fibrous-porous nonwoven material "POROIL" produced by Catensa Ltd.	–	–	20
Nonwoven needle-punched PFM of Toytingen Co. containing		110	7.6
PAN fibers, 72	20		
Cellulose fibers, 22	70–100		
Lignocellulose particles, 6	1000		

application is elimination of the emergency aftereffects in petroleum spills. PFM are made as rolls, hoses, or floating cushions and remove petroleum from the surface of water to protect the banks of ponds and accumulation of petroleum spilled in accidents in industry or transport [4].

A high degree of adsorption of emulsified petroleum oil (60–90%) is observed in filtering oily water through melt-blown PFM, even under tough dynamic regimes [3]. The filtration efficiency is all other conditions being equal, inversely proportional to ρ and d_f and depends on the lipophilic properties of the polymeric fibers (Fig. 9.3).

The behavior of $E(d_f)$ and $E(\rho)$ dependencies reflects the fact that oil products are captured n as a result of their adsorption on fiber surfaces, and also by capture of coagulating oil droplet in the material pores. This agrees well with the theoretical understanding of the filtration mechanisms for oil emulsions in fibrous materials [6].

9.3 Complex-Forming PFM

This type of material is characterized by the presence of particles of functional components (FC) fixed on fibers that can interact with heavy metal

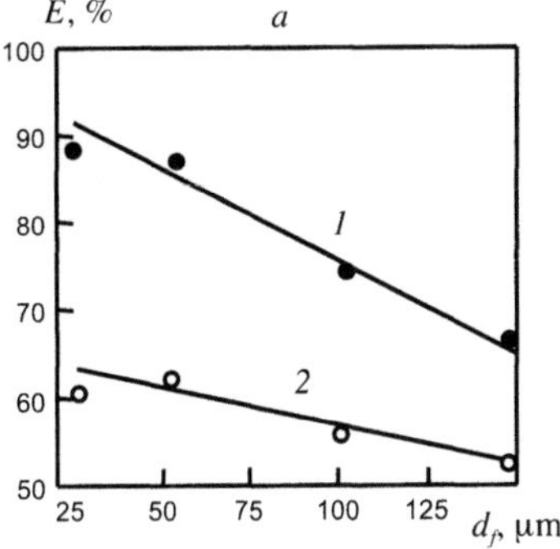
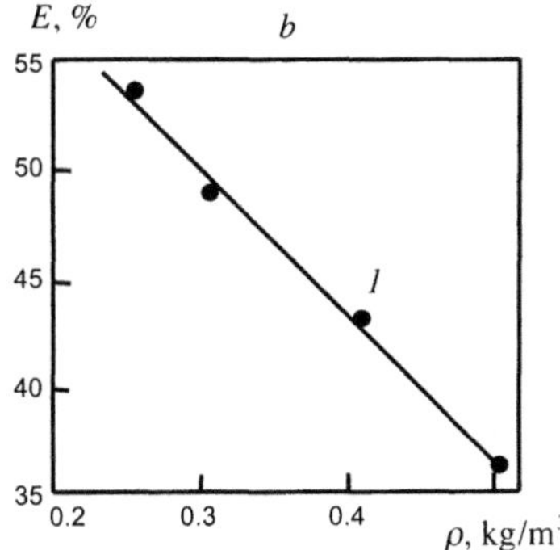

Fig. 9.3. The filtration efficiency of petroleum oil emulsion in water (200 mg/L) depends on the structural parameters of PFM based on PE (1) and PP (2): (a), $\rho = 350$ kg/m^3; (b), $d_f = 30$ µm. FE thickness: a, 9 mm, b, 3 µm.

ions dissolved in the filtered liquid and form stable insoluble coordination compounds (CC) [7–10].

Polynitrogen heterocyclic compounds (PHC) containing one or a few tetrazole (Tet) cycles in a molecule were used as FC (Table 9.3). These PHC are apt to behave as polydentite bridge or chelate ligands during complexing reactions with metal ions [11–15]. In addition to, due to their high thermal resistance, they proved suitable for processing with thermoplastic melts.

$PhTet$, $BTet$, and $TTet$ were synthesized at St. Petersburg's State Technological Institute. They are white crystalline powders with needle-like particles 100–150 µm long and 5–10 µm in cross section.

Filter materials were obtained by melt blowing using _ worm extruder with a special spray head [9]. The fiber melt was sprayed with dry compressed air at 80°C. Tetrazole powders were fed continuously by a specially designed device. Isolated by a tubular jacket, flow was supplied to a continuous band where a fibrous mass was deposited. Upon cooling to LDPE solidification, the latter was separated from the band. The extrusion temperature and aerodynamic regimes of the gas-polymer-tetrazole spray mixture were varied to control the fiber diameter of the polymer matrix and the degree to which it was filled by tetrazole. The matrix porosity was monitored by rolling down the fibrous mass deposited on the substrate with a special roller.

Complexing agents poorly or completely insoluble in water would be appropriate for use as components of PFM intended for capturing metal ions from wastewater. $PhTet$, $Btet$, and $TTet$ meet this requirement. $PhTet$ solubility in H_2O at 20°C is only 0.06%; $BTet$ and $Ttet$ are even less soluble. It should be, of course, borne in mind that the solubility of these compounds (like NH acids) might increase in alkaline media.

The results of complexometric analyses [10] have shown that the static sorptive capacities (SC$_{Cu}$) of $PhTet$, $BTet$, and $TTet$ for the ions of Cu (II)

Table 9.3. Heterocyclic compounds in tetrazole series

Designation and structural formula	Molecular mass	Melting point ($^\circ$C)
5-Phenyl-tetrazole (*PhTet*)	146	213
Bis(β-tetrazole-2-ilethyl) ether (*BTet*)	210	200
Tris(β-tetrazole-5-ilethyl)nitromethane (*TTet*)	349	211

are 6.4, 10.5, and 5.7 mg-eq/g, respectively. PFM containing 15–20% of the PHC cited, this index is about 0.5–2 mg-eq/g.

The values of SC_{Cu} obtained make it possible to suppose that *PhTet* forms coordination compounds with Cu (II) in neutral aqueous solutions in a 2:1 stoichiometric ratio, whereas *BTet* and *TTet* are in a 1:1 ratio. This is supported by elemental analysis of CC (Table 9.4) and agrees well with reported data on analogous PHC complexing with metals [11, 13, 16].

A distinguishing feature of *PhTet*, *BTet*, and *TTet* interaction with Cu (II) during experiments was its heterogeneous behavior limited kinetically by diffusive processes at the solid (PHC)-liquid ($CuSO_4$ solution) interface.

Comparison of the IR spectra of PHC under study and their coordination compounds with Cu (II) yields the following conclusions:

1. Hydrated (or aqueous) complexes are formed upon PHC coordination: a strong wide band of ν(O-H$_{assoc}$) is present at 3468 cm^{-1} (Cu/*BTet*) or 3422 cm^{-1} (Cu/*TTet*), band δ(HOH) is near 1640 cm^{-1}, and adsorption

Table 9.4. Element analytic data of coordination compounds of PHC with $CuSO_4$ (H_2O, 20°C)

Coordination compound	Content of elements (%)					
	Experiment			Calculations		
	C	H	N	C	H	N
$Cu(C_6H_5CN_4)_2 \cdot 1.5H_2O$	44.16	3.33	29.36	43.69	3.64	29.13
$Cu(C_6H_8N_8O) \cdot 1.5H_2O$	25.19	3.72	37.42	24.12	3.68	37.52
$Cu(C_{10}H_3N_{13}O_2) \cdot 1.5H_2O$	27.63	4.84	40.7	27.43	3.66	41.6
$Cu(C_{10}H_3N_{13}O_2) \cdot 2H_2O$	–	–	–	26.88	3.81	40.7

takes place in the 500–700 cm^{-1} region due to liberation oscillations of water molecules in the solid phase ρ (HOH) [17].

2. Heterocycles take part in the formation of coordination links: upon PHC coordination, the whole of the IR spectrum varies responding to valent and deformation vibrations of ring links (within 500-1600 cm^{-1}); spectral bands are present in low-frequency, some of which can belong to valent vibrations of coordination links ν(Cu-N) [13, 16-18].

3. Deprotonated cycles are brought into coordination with the metal: outer sphere anions SO_4^{2-} are absent in CC according to IRS and elemental analysis; within the vibration spectra of CC, the group of 3100–2200 cm^{-1} bands which is found in the initial spectra of PHC is not discerned, and it is related to valent vibrations of the iminogroup of the ring ν(N-H) and ν(N-H$_{assoc}$) in a system of associative cycles interlinked through hydrogen bonds [19].

4. *PhTet* probably forms bridge type, and *BTet* and *TTet* form chelate types of CC [16, 21].

A plain model was made of a composite complexing sorbent to clarify the mechanism of metal ion interception by PFM modified by PHC. It is simply a PE film filled with tetrazole particles.

The main problem to be solved was whether the mechanism of spatial and electron interaction between the ligand and the metal is preserved when complexing proceeds *in situ*, on the surface, or in the polymer bulk.

With this aim, the physicochemical properties and crystalline structure of two types of CC were studied: (1) that obtained as a result of mixing aqueous solutions of *Tet* and $FeCl_2$ or $CuNO_3$ (method A) and (2) that forming on the surface of PE film filled with *Tet* on exposure in aqueous solution of the salts named (method B). It was shown earlier that by introducing *Tet* (15–40 mass%) into the polymer binder followed by hot pressing at $T = 135$–140°C, the respective powder compositions does not give rise to any chemical bonds between the binder (PE) and the filler (*Tet*), and the latter did not lose its individual features.

Elemental analysis as well as IRS, have shown that, independently of the method of synthesis (A or B), CC of the same composition are formed, i.e., $Fe(CHN_4) \cdot 2H_2O$ or $Cu(CHN_4) \cdot 1.5H_2O$. The difference is not observed in the molecular structure of the complex but only in the rate of complex formation. The rate drops abruptly upon transfer from the reaction of reagents dissolved in water (A) to the heterogeneous reaction (B) which is limited by diffusion processes. The latter, it was noted, accelerates with rising temperature in the medium in which the polymer matrix filled with complexing agent is highly permeable.

By comparing the IR spectra of tetrazole, its sodium salt and coordination compounds *Tet* with Fe (II) and Cu (II) it has been established that the role of the ligand in the CC under investigation is played by the tetrazolate anion. *Tet* is a weak N-H acid tending to enter into coordination linkage with ions of transient metals in deprotonated form CHN_4^- (Tet^-) at complexing in a medium at pH > 2 [9].

The IR spectra of sodium tetrazolate and CC containing a deprotonated cycle are devoid, in contrast to the *Tet* spectrum, of the bands corresponding to the link ν(H-N) and associative links ν(N-H...N) at 3050 cm^{-1} and 2300–2900 cm^{-1}, as well as off-plane deformation vibrations δ(N-H) at 935 cm^{-1}. A typical initial *Tet* band at 1520 cm^{-1} induced mainly by valent oscillations of C=N and N=N ring fragments is also absent. The power field of the *Tet* molecule changes, probably, so that the differentiation of valent vibration bands of the cycle at 1520 and 1450 cm^{-1} disappears.

When *Tet* interacts with Fe (II) and Cu (II), coordination takes place with the atoms of the heterocycle nitrogen. This is proved by the emergence of valent vibration bands in the IR spectra of the CC corresponding to Fe-N and Cu-N links in the low-frequency region and a number of intensive bands within the 1315–1365 cm^{-1} region. The latter are not observed either in the *Tet* or sodium tetrazolate spectra and are probably the constituent frequencies of ν(Me-N) and ν, δ(ring) at 1015–1060 cm^{-1}. Furthermore, the N1s band in the X-ray electronic spectra corresponding to cyclic nitrogen atoms expands and changes a lot upon *Tet* transfer to a CC. This is a evidence that nitrogen atoms participate in coordination.

If we consider the peculiarities of the electronic structure of tetrazole and its anion, it becomes obvious that coordination of the cycle with Cu (II) by carbon atom is highly unlikely. This is proved by CC spectroscopic data.

Together with *Tet*, water molecules could be involved into coordination with ions of transient metals. So, a band appears at 415–420 cm^{-1} in the IR spectra of CC which can be related to valent vibrations ν(Me-O) of water linked with a metal ion through a partially covalent bond [17]. However, it is improbable that any clear boundary can be drown between the states of coordinated and crystalline water in the CC according to IR spectral data.

Experimental results agree with suppositions formulated in the literature on the bidentate-bridge function of the tetrazole ligand [12, 13]. According to quantum-chemical notions of the electronic structure of the Tet molecule and its anion [21], atoms of N_1, N_2, or the N_4 ring are said to be the most probable donor centers for coordination with a metal. Probably, each Tet cycle joins by two nitrogen atoms (by the main and side valences) to different Me (II) ions, so that the CC formed has a polynuclear structure. One of the links (through the N_1 pyrrole atom of the cycle) bears a chiefly ionic character. The other that has the donor-acceptor type through a pyridine atom (N_2 or N_4) of cycle is covalent. The (Me-N) band at 250–255 cm^{-1} in the IR spectra of the CC can be, most probably, referred to as an ion type link, whereas ν(Fe-N) frequencies at 390 cm^{-1} and ν(Cu-N) at 325 cm^{-1} are related to a coordination link.

X-ray diffraction analyses have visualized that copper (II) tetrazolate possesses a crystalline structure and iron (II) tetrazolate has a partially crystalline structure. The data on stoichiometry of the CC and on a bidentate character of the tetrazole ligand provide grounds for a supposition about either an octahedral structure (with at least one water molecule participating in the coordination) or a tetrahedral (if water is only crystalline) structure of the coordination node (respectively, MeN_4O_2 or MeN_4). The insolubility of substances in water and organic solvents speaks in favor of a polynuclear structure of the CC molecule.

From the preceding procedure, it follows that introduction of a complexing functional component into the polymer matrix, in particular into a PFM, does not influence on the CC molecular structure formed by its interaction with metal ions, but facilitates diffusive control of the process. So, based on results of physicochemical investigations the following principal factors affecting the complexing properties of composite PFM can be singled out: (1) complexing ability of PHC under study; (2) their concentration in a FM; and (3) system parameters determining the kinetics of diffusion processes at the composite sorbent-liquid interface. It is also to be taken into account that, cation-exchange groups (–COOH and other) are found on the surfaces of oxidized polymer fibers of PFM that canto participate in metal ion interception from filtered media.

9.4 Adsorptive-Microbicidal PFM

When PFM incorporate particles of porous adsorbents fixed on fibers, they can be easily impregnated with microbicidal preparations (either natural or synthetic). Such treated PFM can simultaneously fulfill the functions of filtration and adsorptive cleaning, as well as disinfection of gases and fluids.

Sometimes, sorptive modifiers have bactericidal activity of their own. This feature is ascribed, particularly, to some of the tetrazole compounds. Microbiological tests were conducted on PFM with fibers plasticized by oil extracts of birch leaves (BLE) or coniferous needles (CNE) as well as fibers containing adhesively bonded particles of tetrazole or AC impregnated by these

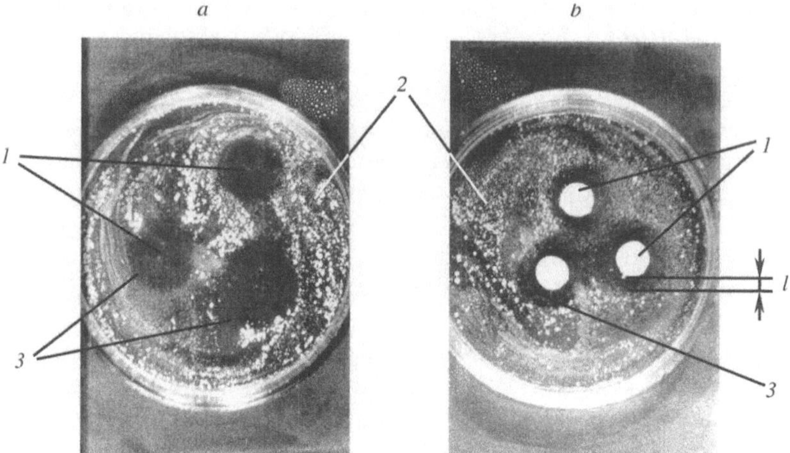

Fig. 9.4. Photos of Petri dishes with an agar culture medium and samples of *PhTet* (**a**) and PFM modified by *PhTet* (**b**) after 48 h of *Sarcina flavus* bacteria cultivation: *1*, samples; *2*, colony of microbial cells; *3*, zones of microbial growth suppression; *l*, zone *3* width

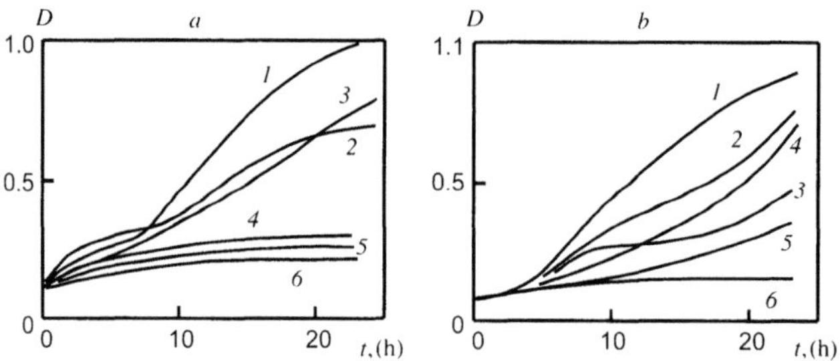

Fig. 9.5. Kinetics of optical density (D) variation of *Sarcina flavus* suspensions (**a**) and *Pseudomonas fluorescens* (**b**) in a aqueous culture medium: 1, check test; 2–6, upon introducing different composition of PFM into suspensions: 2, pure PE; 3, PE + AC; 4 and 5, PE and PE + AC impregnated by coniferous needle extract; 6, PE + PhTet

extracts. All they display a distinctly expressed antibacterial effect in respect to coccus microflora, gram-negative sporeless, and gram-positive spore bacteria (Fig. 9.4, Table 9.5).

The effect of biocidic PFM on the growth dynamics of test-microbes in aqueous culture media consists of prolonging the growth-adaptation phase (lag-phase) and in fact full suppression of the logarithmic growth phase (Fig. 9.5) [10, 12].

It has been found that even nonmodified melt-blown PFM on a PE base intercept bacterial contaminants efficiently, especially coccus microflora.

Table 9.5. Zone width (l) of microorganism suppression around microbicidal FC and PFM samples modified by these FC (30°C, agar culture medium)

Sample type	Sample composition	Cultivation time (h)	Zone width l (mm) in cultivation of microorganisms		
			Sarcina flavus	*Proteus vulgaris*	*Bacillus subtilis*
FC	CNE[a]	48	5–9	3–7	7–9
	BLE[b]	48	2	2	1–3
	AC + CNE[a]	24	13–14	5–7	12–13
		72	6–8	4–5	6–7
	PhTet	48	15–20	5–7	10–15
PFM	PE + CNE[b]	24	5–7	2–3	2–4
		72	2–3	1–2	2–4
	PE + BLE[b]	24	4–5	2–3	3–4
		72	2–3	1–2	2–4
	PA + CNE[b]	24	4–5	2–4	4–5
		72	3	2–3	3–5
	PA + BLE[b]	24	5–6	3	4
		72	3	2–3	4
	PE + AC + CNE[a]	48	–	2–3	6–8
	PE + *PhTet*	48	10–12	2–3	7–10

[a]alcohol or alcohol-acetone extracts.
[b]extracts in Vaseline oil.

Investigations were carried out at the Department of Prophylactic Toxicology of the Hygiene and Toxicology Center of the Belarus. Samples based on running water contaminated with microbiological admixtures were used for testing. The greatest filtration efficiency was observed with microorganisms: 86% for cleaning from *Enterobacteriacae*, 78% from coccus microflora, and 30–38% from spore bacteria.

10. PFM as Carriers of Microorganisms

A present trend in industrial technologies is the growing role of catalytic biotechnological systems, in particular, systems for biologically cleaning air and water by employing of biologically active polymer materials (BAPM). The functional mechanisms of these materials (where a high molecular weight matrix is the carrier of live cells of microorganisms, plants, animals, or men) are governed by the metabolism of immobilized cell cultures under given conditions. The rapid widening of the BAPM range in recent decades can be attributed to the strengthening of the biotechnology sectors in a number of the vital domains of economy, including fine organic synthesis, the pharmaceutical and food industries, medicine, agriculture, and industrial ecology [1].

10.1 Biofilters with Polymer Fibrous Biomass Carriers

Biofilters can be used for efficient and simple cleaning to carry out thorough biological utilization of sewage contamination (through biological conversion into nontoxic forms using microorganisms immobilized on carriers) [2–5]. Aerobic film biofilters with immobile packing (microorganism carriers) are one of the cheapest and most reliable types of such filters. Microbiological degradation of admixtures found in sewage cleaned through biofilters simulates the natural biochemical phenomena of "self-cleaning" in nature. Under optimum regimes of biological cleaning, extremely high levels of adsorption and bioconversion of wide spectrum organic and inorganic contaminants are reached.

The key to successful biocleaning is the choice of active strains of microorganisms or their consortiums, which are first and quick in metabolism in the trophic chain of biologically converting contaminants. The efficiency and reliability of biofilter operations are chiefly determined by the characteristics of the biomass carrier.

Contradictory demands are imposed on the carrier because, on one hand, it should be permeable to the air stream and the medium being filtered and, on the other hand, should keep biomass and protect it against mechanical, aero- or hydrodynamic effects; buffer extra loads for abrupt changes in sewage composition; and provide for maximum cleaning of contaminants. From this standpoint, the most significant parameters of the carrier are the following: the surface developed, the porosity, satisfactory mechanical strength and chemical stability, biocompatibility; that is, nontoxicity to microbial cells and the ability for provide of stable vital activity of microbial associations [5, 6]. Also of great importance are the technical and economic characteristics of the carrier, namely longevity, resistance to heat sterilization, light weight and shape stability, highly efficient production technique, and low cost.

So far, the selection of a carrier for certain industrial biocleaning cycles has been based on considerations of available conventional and cheap materials. Nevertheless, known carriers, both inorganic (sand, crushed stone, gravel, glass, brick, slag, coke, claydite, zeolite, activated coal) and organic natural (shavings, bark, sawdust, peat) as well as artificial and synthetic polymers (cellulose acetate, PE, PP, polyurethane foam, PVC, PS, PVOH, and others) often do not meet the complex requirements cited. Most promising in a number of parameters (including biocompatibility), polymer and composite carriers are today on a par in cost with typical structural and building materials employed for this purpose. Note that most polymer carriers (smooth and crimped films, strips, rods, broaches, rings, tubes, discs, seats, cellular structures of polymer sheets, etc. [7–12] have weakly developed surfaces (50 to 300 m^2/m^3). A layer of foam polymer granules [13] is poorly permeable and is blocked in time. Fine-grain polymers, including gel-formed carriers can to operate only in conditions of fluidization [12, 14]. More suitable for their sorption and other service characteristics, familiar types of fibrous woven and nonwoven carriers [15–17] are not stable in shape. To impart structural permanence to such materials, they are reinforced by frames that make carrier production and operation complicated and impose additional mechanical loads on the bioreactor.

A refined melt blowing procedure has united the production and modification processes for shape-stable fibrous polymer carriers (FPC) into a single production cycle which adds certain functional properties that make them biocompatible with microorganisms and regulates their activity [18, 19].

FPC were formed into cylindrical elements (rings) based on PA, PP, or HDPE with a mean diameter of 40 mm, height 30 mm, and thickness 5 mm. Their structural parameters were varied across a wide range, including d_f from 10 to 140 µm and ρ from 100 to 400 kg/m^3.

Optimal shape and dimensions were chosen as a result of the analysis of known charging systems used in chemical or biological catalysis [3, 20]. When optimizing the thickness of FPC walls, it was taken into account that the maximum depth of aerobic microorganism film ingrowth did not exceed 2–4 mm under conditions of conventional metabolism [11].

A poured layer of FPC of a chosen shape and dimensions ensures the most balanced and rapid mass exchange in the bioreactor bulk, good permeability for flows in every direction, uniform distribution of mechanical loads on bearing structures, and simplifies carrier substitution and regeneration [21].

The structural parameters of FPC were optimized according to a complex of the following criteria: the mechanical strength and shape stability of the fibrous elements, the porosity and specific surface of the carrier developed to ensure perfect air and water penetrability, as well as sorptive capacity with respect to the microorganisms and admixtures in the filtered media. It has been found that FPC, where $\rho = 200$–250 kg/m^3 and $d_\mathrm{f} = 40$–50 μm for PP or 70–90 μm for PA, correspond best to these requirements. FPC on an LDPE base are less preferable due to their low heat stability under conditions of regeneration by steam sterilization.

Optimum FPC structures can be characterized by rather high breaking stress values under tension (2–3 MPa) and limiting loads under ring element compression (40–120 N) which is especially important in operating industrial bioreactors having a thick filling layer. The specific filling area S_sp in FPC reaches $(7 - 11) \cdot 10^3$ m^2/m^3 with a filling ratio of the biofilter by a fibrous material equal to 0.5. This is much better than the conventional filling ($S_\mathrm{sp} = 2 \cdot 10^2$–$2 \cdot 10^3$ m^2/m^3). The pressure differential Δp in water filtration through FPC walls is below 1 kPa, which is one to two orders of magnitude less than the corresponding values typical of common carriers. Among the advantages of FPC based on melt-blown materials are also low filling mass (~ 100–120 kg/m^3), chemical and biological inertness, and broad opportunities for imparting additional useful functional properties to the carrier material (sorptive, electret, magnetic, etc.)

FPC performance in biofilters has been evaluated by the following parameters:

- specific sorption of complex microorganism cultures per unit mass (a_m), unit geometrical surface (a_s) and unit carrier volume (a_v);
- comparative filtration efficiency (E) of sewage by biofilters filled with FPC under study and conventional (claydite) carriers;
- time for the biofilter to reach steady-state operation regime from initial start-up and upon the prolonged absence of loading.

Indexes a_m, a_s, and a_v were estimated by a standard microbiological method that judges by the difference between the carrier mass before and after immobilization of reference microorganism culture (*Bacillus cereus*, *Aeromonas spec.*, and *Pseudomonas spec.*).

In Table 10.1, specific sorption indexes are shown for complex microorganism cultures on differently structured FPC samples in contrast to typical carriers. As typical carriers, we used Rashig rings of oxide ceramics with a 30 mm mean diameter, 5–50 μm (R-1) and 200–500 μm (R-2) pore size; 20 mm diameter polyurethane foam (PUF) granules; and 20–40 mm fraction of claydite spherical grains. It is evident that FPC surpasses typical microorganism carriers in sorption parameters. The time for optimum structure and composition of FPC for wetting by a culture medium is from 5 to 30 min. A high degree of oxidation of the polymer in the fiber surface layer, it has been noticed, promotes wetting.

Aerobic biological cleaning of wastewaters has been investigated using pilot biofilters (Fig. 10.1) developed at the State Scientific Center for

Table 10.1. Specific sorptive capacity of different carriers for association of the microorganisms *Bacillus cereus*, *Aeromonas spec.*, and *Pseudomonas spec.*

Carriers		a_m (mg/g)	a_s (mg/cm^2)	a_v (mg/cm^3)
Typical	R-1	0.6 ± 0.2	0.06 ± 0.02	0.9 ± 0.3
	R-2	1.6 ± 0.5	0.14 ± 0.03	1.1 ± 0.2
	PUF	2.6 ± 1.3	0.14 ± 0.05	1.1 ± 0.4
	Claydite	4.4 ± 0.6	0.28 ± 0.04	1.3 ± 0.2
FPC	PP-1[a] ($d_\mathrm{f} = 10$ μm, $\rho = 130$ kg/m^3)	8.5 ± 1.4	0.29 ± 0.07	1.6 ± 0.4
	PP-2 ($d_\mathrm{f} = 50$ μm, $\rho = 220$ kg/m^3)	7.6 ± 0.4	0.42 ± 0.03	1.7 ± 0.1
	PA[a] ($d_\mathrm{f} = 25$ μm, $\rho = 250$ kg/m^3)	19.2 ± 4.0	1.20 ± 0.19	4.8 ± 0.6

[a]The material partially swells during sorption.

Applied Microbiology (Russian Federation, Obolensk). The biofilter volume was 5.3 dm^3, the volume and height of the carrier layer were 4.8 dm^3 and 1.75 m, respectively. The specific flow rate of wastes Q (relative to carrier volume) was varied within 0.1–7.3 dm^3/(dm^3·day). The term of regime maintenance under a constant waste rate was, as a rule, 5–14 days.

The filtration efficiency (E, %) of the chemical wastes was estimated by the dichromatic standard method [5] that judges by reduced contamination at the biofilter outlet relative to the inlet value of chemical oxygen demand (COD):

$$E = \frac{\mathrm{COD_{in}} - \mathrm{COD_{out}}}{\mathrm{COD_{in}}} \cdot 100\% \tag{10.1}$$

The efficiency of biological filtration for separate contaminants was estimated by gas chromatograph. The analysis of oil products content (C) in wastes consisted of extraction by chloroform followed by solvent removal and weighing the remainder. The efficiency was calculated by the formula

$$E = \frac{C_\mathrm{in} - C_\mathrm{out}}{C_\mathrm{in}} \cdot 100\%. \tag{10.2}$$

The time for the biofilter to reach steady-state regime was considered period elapsed from actuation till reaching a stable value of $E = 90\%$ and more (for oily wastes).

The FPC efficiency was compared to that of the best typical carriers (claydite) during filtration of wastes from a chemical enterprise, recirculating waters from the cooling network of an atomic power station, and wastes from

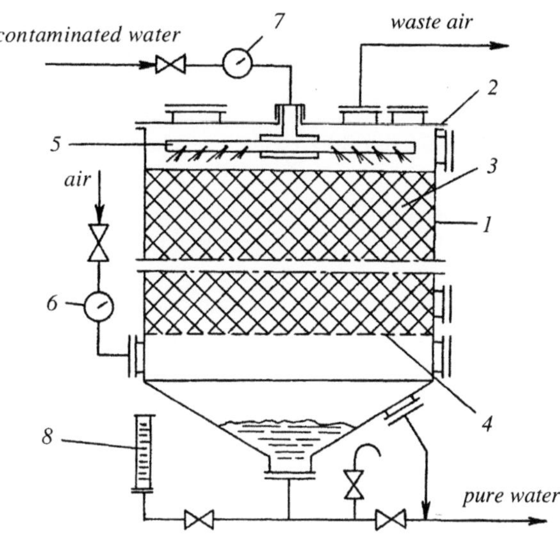

Fig. 10.1. A diagram of a biofilter for aerobically cleaning wastewater: *1*, casing; *2*, cover; *3*, layer of carrier with immobilized mass; *4*, bearing mesh; *5*, spray device; *6*, *7*, flowmeters; *8*, level gage

Table 10.2. Compositions of industrial wastewaters subjected to cleaning by biofilters

Composition No.	Type of waste	COD (g/L)	Type (concentration, mg/L)
			Contaminants
1	Wastes of chemical enterprise	14.1	Isopropynol (770), ethyl hexane (2530), chlorethyl hexil (1840), acetone (900), methyl isobutyl ketone (200), esters (200), sodium chloride (1200), mesityl oxide (100), toluene (50), diacetone alcohol (30)
2	Recirculating water from an atomic power station		Oil-products (10–270)
3	Water purposely contaminated by waste oil		Oils (60–90), heavy metals, antifreeze, and other additions
4	Wastewater from antibiotics production	7–10	Culture liquid residue, antibiotics (benzyl penicillin, oleandomycin), organic solvents (butanol, acetone, benzyl acetate)

antibiotics production. The compositions of the wastes studied are presented in Table 10.2.

It has been found that a biofilter with PP-2 based FPC was highly competitive with a claydite-filled biofilter in purification quality of chemical wastes

(Fig. 10.2). The degree of bioconversion of individual contaminants is within 35–100% independent of the biofilter filling type. However, at start-up, the biofilter with FPC operates better (fresh biofilm).

The time for reaching a steady-state regime in purifying wastewater contaminated by oil products is reduced from 30 to 14 days when the claydite biomass carrier in the biofilter is substituted by the FPC carrier. Aside from this, FPC provides more efficient biocleaning at extensive flow rates, considerable content of oil products in the inlet, and the presence of toxic substances in the wastewaters (Fig. 10.3).

The cleaning of oily wastewater has been modeled under conditions where temporary interruption or appreciable reduction of the waste supply takes place while the biofilm viability is supported by recycled moistening of the carrier or by a small flow through the reactor of water with relatively low oil (below 10 mg/L) and biogenous additive content. After 2 months of operation while keeping the biofilm viable, a rigid filtration regime was upset by an abrupt daily increment of highly contaminated waste flow rate for 5 days. In such conditions an FPC-filled biofilter quickly reaches and shows high stable filtration indexes in contrast to that filled by claydite (Fig. 10.4).

The improved biofiltration efficiency through a FPC-based filter is probably related to the affinity of its carrier with the biomass (Table 10.1) as well as favorable conditions for biofilm growth in fast mass exchange processes due to the perfect air and water permeability of fibrous elements.

The advantages of polyolefin based FPC in systems for biological filtration of oily wastewaters depend significantly on the ability of lipophilic polymer fibers to adsorb great amounts of oil products thus providing free access by the

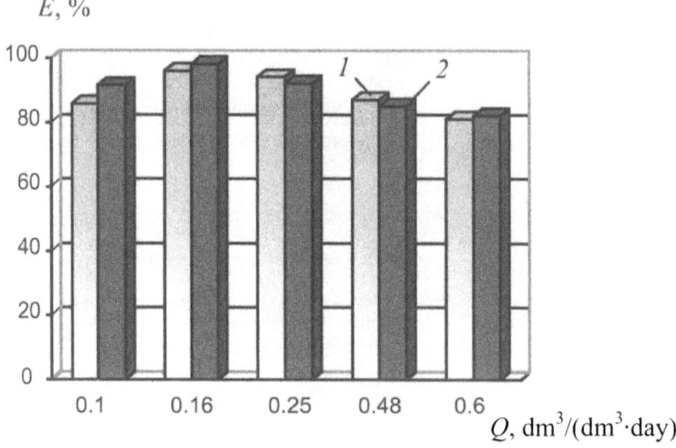

Fig. 10.2. Efficiency (E) of biological cleaning of wastewater from a chemical factory (composition no. 1 in Table 9.2) during one stage by biofilters with claydite (1) and FPC (PP-2) (2) carriers depending on the specific flow rate of wastes (Q). Working association of microorganisms *Bacillus cereus, Aeromonas spec.,* and *Pseudomonas spec.*

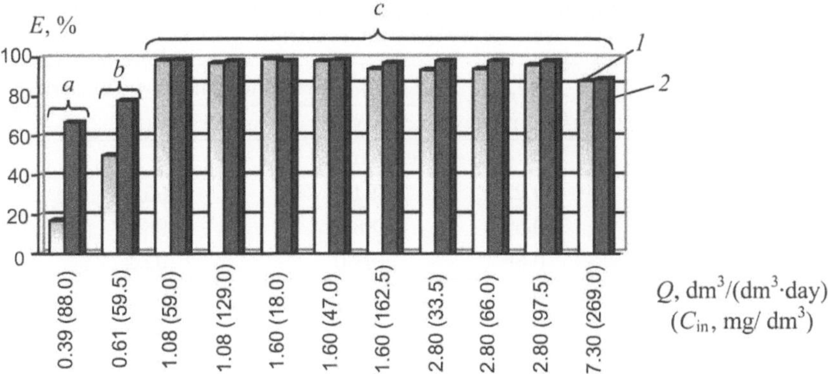

Fig. 10.3. Efficiency (E) of biological cleaning of wastewater contaminated by oil product steady state process regimes versus specific flow rate (Q), inlet concentration of petroleum products (C_in), and the presence of toxic substances. Contaminant: (**a**) waste toxic oils with admixtures of antifreeze and ions of heavy metals; (**b**) transmission oils with additives; (**c**) oil products of APS cooling waters. Microorganism carriers: *1*, claydite; *2*, FPC (PP-2). Working association of microorganisms *Acinetobacter spec.*, *Pseudomonas spec.*, and *Mycobacterium flavescens*

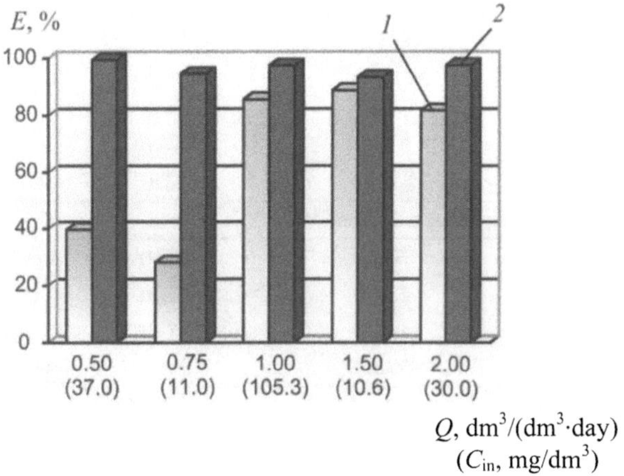

Fig. 10.4. Efficiency (E) of biological cleaning of wastewater contaminated by used transmission oils with an abrupt change of the biofilter regime from "biofilm maintenance" to "operation." Carriers: *1*, claydite; *2*, FPC (PP-2)

biofilm to the substrate. In Chapter 9, the high adsorptive capacity of polymer fibrous materials on PP and PE bases for oil products has been described in detail (see Table 9.2). The amount of oil products isolated from FPC samples during extended operation in a biofilter is, nevertheless, negligible (5–8 mg/g) because the contaminant is converted by vital activity of the microbial association.

Table 10.3. Biocleaning efficiency of antibiotics production wastewater following a one-stage continuous flow scheme without recirculation

Biocleaning cycle, COD_{in}, g/L	Q, [dm^3/(dm^3·day)]	$E(\%)$ for biofilter filled with carrier:	
		Claydite	FPC (PP-2)
Steady-state regime[a]			
I	0.28	94.0	95.2
8.3	0.50	88.6	98.8
	0.86	88.0	92.2
II	0.42	82.0	94.4
8.0	0.99	58.1	89.4
	0.56	83.8	91.2
	2.65	38.1	63.1
III	1.04	75.6	94.4
8.0	1.67	88.1	88.1
	2.14	47.5	93.8
IV	1.52	70.1	91.0
7.2	2.06	53.5	86.1
	3.13	38.9	58.3
	4.01	41.7	44.4
V	1.07	70.3	45.9
7.4	2.08	56.8	70.3
	3.07	40.5	67.6
	4.62	51.3	37.8
Transient regime after a 3-week stop[b]			
VI	0.51–0.56	70.0	88.0
10	0.74–1.54	42.0	54.0
(wastes with	1.03–1.89	48.0	76.0
suspension)	1.63–2.65	56.0	50.0

[a] Q was gradually increased daily.
[b] Q was increased for a day ("acute" test).

Moreover, FPC shows the ability to damp toxicological effects on the biofilm. This was proved during biological purification of wastes contaminated by toxic oils (Fig. 10.3a,b) and was especially evident in filtering wastes from antibiotics production. When considerable amounts of organic solvents and antibiotics toxic to microorganisms are present in wastewater, the process is insufficiently stable in both the transient and steady-state regimes (Table 10.3). But as a whole, biofilters with FPC operate more efficiently and reliably than those with claydite filling.

Thus, FPC provide highly efficient operation of biofilters and outdo typical microorganism carriers in all key criteria. For example, FPC which have been used for 4 years in pilot biofilters in biological cleaning of heavy duty wastewater and have undergone multiple charging, discharging, and vapor sterilization, have not shown any signs of breakdown or impairment in technical parameters. Long experience in the operation of pilot biofilters with FPC as well as analysis of accumulated statistical data have provided the basis for scaling the process up to the commercial level of biological cleaning.

Furthermore, prospects of improving the service characteristics of the proposed FPC are connected with physicochemical methods of polymer modification during the melt blowing process, namely, electric polarization, fiber filling with magnetic and conducting substances, fixing of selective dispersed adsorbents on the fiber surfaces, and impregnation of culture media for microorganisms into the fiber surface layer. In particular, adsorptive concentration of admixtures on the carrier occurs when filtering through a composite FPC containing highly porous adsorbents (AC and other) which alleviates microbial degradation.

FPC can acquire new properties as a result of their chemical structure modification for affinity to microbial cell membranes. They can generate systems controlling the biological activity of immobilized microorganisms and its feedback for processes of biological cleaning. Qualitatively new potentialities of FPC with controllable microorganism activity allow us to relate them to the latest generation of biocatalysts and bioadsorbents governing progressive trends in engineering ecology.

10.2 Effect of Magnetic Fields on the Growth Processes of Microorganisms

Taking into account information on the orienting and stimulating influence of a MF on microbial cells [22–24], it would be sound to suppose that the application of magnetic-filled PFM as a microorganism carrier would assist in biofilter operation. Thanks to a MF it is possible to control the distribution and enhance the retention of microorganisms on a magnetic carrier, to intensify their growth, and to magnify their biodestructive ability. In particular, the rapid growth of a biofilm is important for prompt biofilter actuation. In our view, among promising microorganism carriers the best are melt-blown PFM to be modified by dispersed magnetosolid ferrites.

In [25], the regularities of a weak MF effect of magnetic PFM have been studied, as well as solid and film samples of magnetic plastics in the processes of microorganism test-culture growth in solid and liquid culture media. The degree of magnetoplastics filling by magnetic was varied within the $C = 1$–90 mass% range. Ferrite-filled cylindrical samples magnetized by pulse MF in axial and radial directions displayed surface magnetic induction of about $B_\mathrm{r} = 0.05$–27.6 mT which depended upon magnetic filler concentration, material texture, and magnetizing regimes.

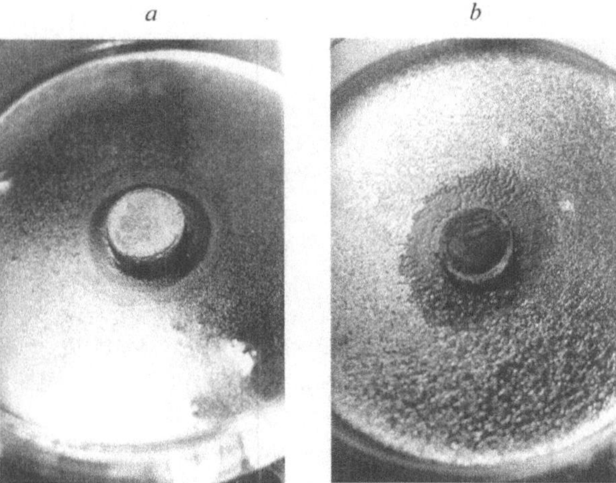

Fig. 10.5. Ring zones of *Ps* stimulated cell growth on meat–peptone agar around cylindrical samples of magnetoplastics: (**a**) radial magnetization, $B_r = 5.75$ mT; (**b**) axial magnetization, the north pole is in the agar, $B_r = 0.25$ mT

The MF of magnetoplastics, it was found, exerts intense action on the growth processes of *Pseudomonas fluorescens* (*Ps*), *Staphylococcus* (*St*), and *Aspergillus niger* (*As*) test microbes on solid culture media. Cylindrical samples of fibrous and solid magnetoplastics magnetized in the radial direction were placed in Petri dishes with meat–peptone agar. After 48 h of *Ps* bacterial cultivation, ring zones of stimulated microbial cell growth interpolated with those of retardation were observed around the samples. Their width and location relative to the samples, it was found, depend on the B_r value. The most favorable conditions for accelerated growth of *Ps* arose in a weak MF of fibrous and solid magnetoplastics with low ferrite concentration ($B_r = 0.1$–0.5 mT). In this case, growth zones with $l = 7$–12 mm wide were located along the sample perimeter. At $B_r = 1.6$–5.8 mT zones of stimulated growth shifted some distance from the MF source, and zones of retarded microbial cell growth started to nucleate around the magnetoplastics (Fig. 10.5a). Furthermore, at $B_r = 27.6$ mT, ring zones of stimulated growth were found again close to the sample vicinity. Using blocks formed from magnetized ferrite powders without polymer binder, only 3-5 mm wide zones of *Ps* growth retardation were detected. Presumably, ferrite particle encapsulation by the polymer improves the biocompatibility of magnetic materials with microbial cells.

Microscopic analysis of microorganism samples cultivated for 6 days proved that large viable cells typical of *Ps* were present in the ring zones of stimulated growth. Beyond these zones, polymorphism of microbial cells was recorded which was the sign of the onset of growth attenuation [26].

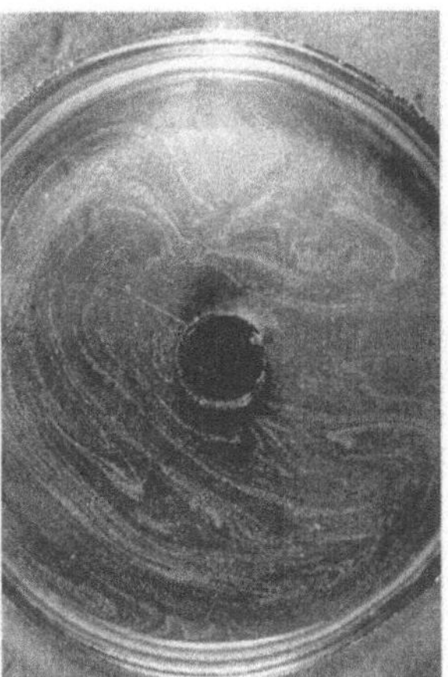

Fig. 10.6. Oriented *Ps* cell growth on an agar culture medium in a MF generated by a magnetoplastic cylindrical sample: axial magnetization $B_r = 2.6$ mT

As can be seen, microbiological testing has shown that MF of a certain energy generated by magnetoplastics positively influences the processes of cellular component replication that lead to accelerated multiplication of *Ps* cells and prolong their metabolic vitality.

Samples of *As* and *St* microorganisms magnetized in radial direction reacted less selectively to the MF effect. Nevertheless, in all cases of MF application, more intense microbial growth was observed on the solid culture medium in contrast to the reference tests in the absence of a MF.

The following phenomena recorded during experiments with samples magnetized in the axial direction are to be noted. The hastened growth of *Ps* test microbes occurred in the vicinity of the north pole of samples immersed in the culture medium (Fig. 10.5b). Near the south pole microbial growth was, vice versa, suppressed. The *St* population grew evenly across the whole agar area in the Petri dish, whereas replication of microbial cells increased in response to augmented MF induction. Under analogous conditions, complex vortical behavior of heaps of *Ps* bacteria cells (Fig. 10.6) or fungi *As* mycelium was detected on the surface of the culture medium. This proves orienting effect of a magnetoplastic induced spatially inhomogeneous MF on the growth processes of the microorganisms mentioned.

The oriented growth of *Ps* bacteria in a spatially inhomogeneous MF can be attributed to their paramagnetic properties. This was supported by EPR spectra of *Ps* cells adsorbed from a liquid phase on PA-based PFM samples (Fig. 10.7). The EPR signal generated by immobilized bacteria evidently belongs to organic radicals that arise due to ion–radical biochemical reactions in the cells [27].

Cylindrical blocks of magnetoplastics magnetized in the axial direction and placed in a Petri dish (shown in Fig. 10.6) generate a MF whose force

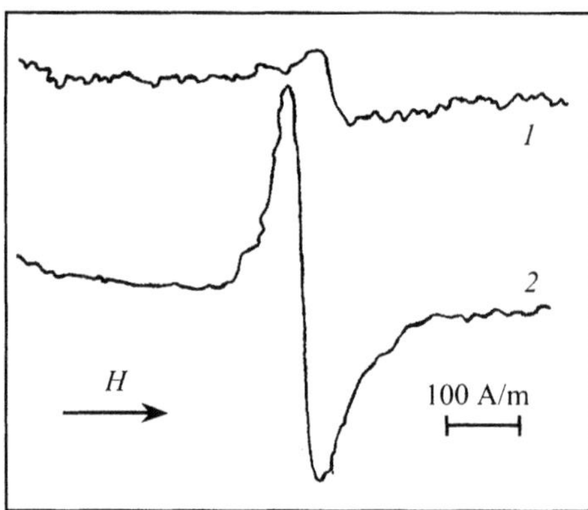

Fig. 10.7. Electron paramagnetic resonance (EPR) spectra of PA-based PFM samples before (1) and after (2) adsorptive immobilization of *Ps* cells from an aqueous culture medium

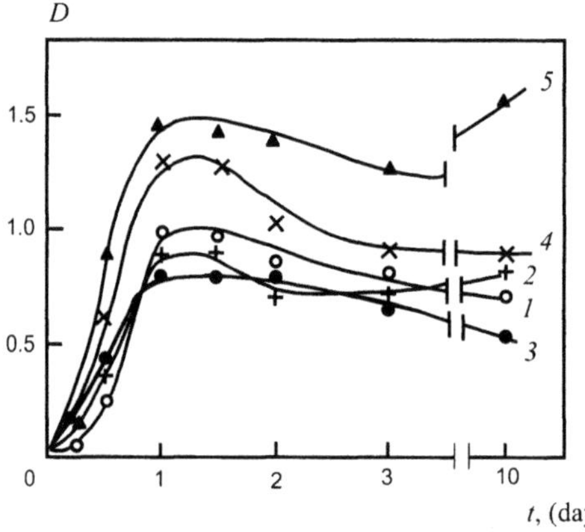

Fig. 10.8. The kinetics of the variation in optical density (D) of *Ps* suspensions in an aqueous culture medium: *1*, reference test; *2–5*, in placing polymer samples into suspensions; *2* and *3*, nonmagnetic PE- and PA-based PFM; *4* and *5*, magnetic PE- and PA-based PFM ($B_\mathrm{r} = 0.1$–$0,3$ mT)

lines are perpendicular to the agar substrate surface. Due to the structural peculiarities of PFM, the field is, however, distributed unevenly in space where intensity splashes alternate with a drop. Under such a field configuration, *Ps* test microbes, it is anticipated, localize in areas where energy levels of electromagnetic effects are most favorable for their vitality.

The biological activation of *Ps* cells in liquid cultural media induced by the MF of magnetoplastics leads to an accelerated phase of logarithmic growth of the bacteria (1.2–2 times) and, consequently, to a biomass increment (Fig. 10.8).

As result, a weak MF generated by magnetoplastics (mainly showing $B_\mathrm{r} = 0.1$–1 mT) was established to exert a stimulating effect on the growth and metabolic processes of microbial cells. The mechanism of the magnetobiological effect probably consists of MF interaction with paramagnetic molecular targets in the cellular structures. They include, e.g., some enzymes that participate in biochemical reactions with electron transfer. These reactions can be involved with the main processes of microbial cell survival (energy supply, breathing) and replication [27–29].

So, available data on the spatial orienting and growth stimulating effects of a MF generated by magnetoplastics on microbial cells suggest that magnetic PFM are a promising tool for use as microorganism carriers in biological filters. The spatial orienting effect of a MF on microbial cells growth can be efficiently employed to control immobilization processes of microorganisms on magnetic carriers and to establish a preset biomass distribution within the filtering layer of a bioreactor.

At present, these ideas are realized in biofilters for wastewater purification in some enterprises in Russia and South Korea.

11. Other Applications of PFM

The service characteristics of melt-blown materials depend strong on the high specific area and nonequilibrium state of the fiber surface layer that has rather high surface energy. The physicochemical activity of this class of materials in various technological, working and natural media has defined their optimum fields of application in industry, medicine, construction, and other important spheres. New means of efficiently cleaning running water and foodstuffs, as well as technological media, wastes, and industrial waste gases are extensively explored now. Disposable melt-blown hygienic and engineering products, novel grades of textile decorative fabric, leatherette, carpeting, and other household articles have become of paramount importance for the modern generation. Melt-blown materials have already occupied a significant place in civil construction successfully solving problems in heat, sound, and waterproofing of buildings and various structures. Today's advances and prospects in medical technique development are also improbable without fibrous materials. A powerful global packaging industry that has expanded during the last half century employs melt-blown products widely, too. A diagram presented in Fig. 11.1 shows the potentialities of melt-blown materials and visualizes the main trends in their use. Following, some characteristics of the group of materials mentioned in the diagram are given.

11.1 Household Uses

These embrace a broad range of melt-blown materials processed into individual articles or components of composite materials and combined articles, including disposables (napkins, towels, kerchiefs, clothes, linings, etc.), textiles for decorative and household use, leatherette, carpets, and soft carpeting for floors.

Melt-blown disposables consist of a wide range of products.

Towels are obtained through a diversity of technological effects on fibrous cloth, which add to them a number of textural peculiarities and attractiveness for consumers. Such towels can

- be fluffy on one side and smooth on the other;
- have two water absorbing surfaces;

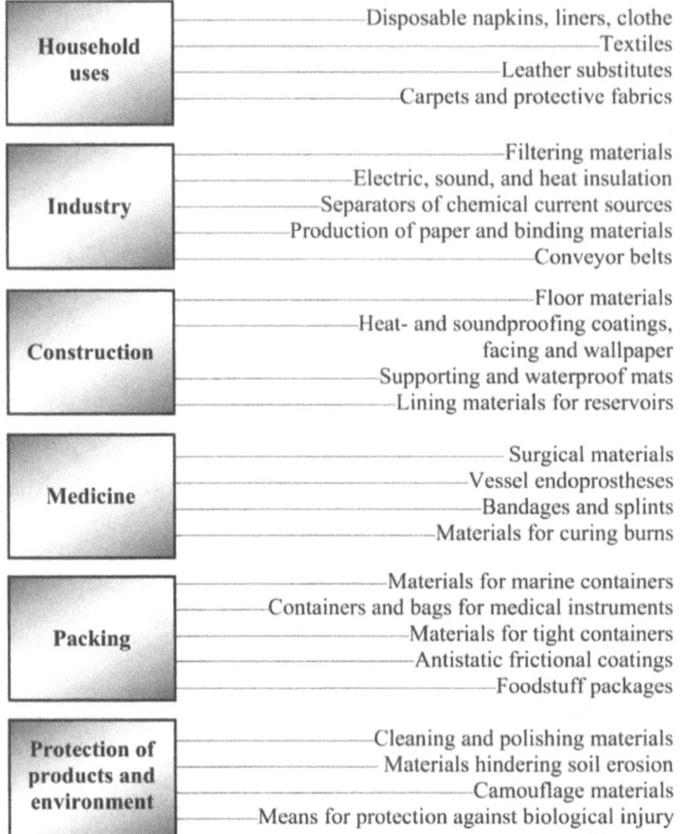

Fig. 11.1. Fields of application of melt-blown materials

- have a surface formed by areas of unlike wettability;
- contain mineral adsorbents;
- be multilayered with the middle absorbing layer of greater mass compared to the outer fluffy layers.

Original techniques have been developed for each of articles named, as well as for equipment and rigging [1].

Hygienic napkins are to meet a combination of rather incompatible and even mutually exclusive requirements. One of most practical problems encountered in their design is how to lower the fiber tendency to crack in bending that leads to loss of shape, skin irritation, and discomfort. To eliminate this disadvantage, it is necessary to introduce lubricating-plasticizing additives into the fiber composition, which prevent the pile from sticking together, impart softness to the product and a feeling of comfort in application. Kimberly–Clark Corporation was the first to design such napkins.

Another way of solving this problem is to increase the deformability of the product. For example, melt-blown polypropylene napkins with an assigned fibrous structure are subjected to extension under optimum technological regimes to 150length without rupture, e.g., by process [2]. They are employed in manufacturing nappy liners and diapers.

Sanitary napkins are most often produced from two and more types of fibers one of which forms a reinforcing network (e.g., the base is PE, and the network is PP). The leading trend in designing multilayer napkins has become inserting absorbing layers within the fibrous structure that provide regulated density to force the capillary flow of absorbed moisture in a preferable direction. This condition is especially important for hygienic napkins whose length far exceeds their width.

Biodegradable articles represent the leading trend in modern production that develops in conditions of environmental global pollution. Melt-blown fibers bonded together by a resin based on cellulose sulfur ester salts are sufficiently strong and resistant to abrasion in the dry state and also when moistened with urine, blood, menstrual fluid, etc. as well. At the same time, they are readily soluble in water and antiseptic solutions, can be flushed down the toilet and do not clog piping. Analogous properties are displayed by melt-blown napkins based on a cationic PU binder. Along with this, a series of products has been elaborated based on fibers bonded by a binder that decomposes in enzyme-containing media so that it is, e.g., a piping effluent. A variant of this solution consists of using pH-sensitive polymer binders, in particular, alkyl-acrylics. Toilet melt-blown towels moistened at usage readily disintegrate in water and easily pass through pipelines.

Disposable clothes whose strength is not impaired by wetting are most often produced from a mixture of twisted cellulose and synthetic filaments mechanically hooked. Disposable materials for clothing are also made by adding polymer binders (acrylic, cellulose) by drying a layer of foamed suspension based on a binder and fiber mixture. Another material production technique also widespread uses a layer of short fibers glued to a melt-blown cloth.

Fabrics for cleaning and polishing floors coated by asphalt, linoleum, tile, polyvinyl chloride plastic, and so on, have a melt-blown base of about $40\,\mathrm{g/m^2}$ specific mass impregnated with a water-based polishing composition. It incorporates solvents, alcohol additions, and polymer emulsions. The composition mass ratio to that of the base is from 3.0:1 to 6.5:1. Multifunctional materials are intended for wiping spilled oils, and drying smooth surfaces using solvents or wet cleaning.

Textile materials for trimming clothes and for domestic use consist of the following categories of melt-blown products.

Fluffy articles are attractive and highly resistant to abrasive wear. In the simplest case, they consist of finefibrous cloth and coarse linen connected by needle punching [3]. As a rule, the product obtained by needle punching two cloths having a pile only on one side and folded by smooth sides loses, with time, the elasticity intrinsic for melt-blown materials. To preserve the shape of such a multilayered material and the finished article at a maximum, a binder is often applied on the facing sides, which is set either by heat or chemicals. Decorative textile products are made of a colored terry fibrous material connected by either needle punching or by a binder with a cloth-substrate of a certain color, e.g., to form an ornament.

Articles with characteristic trimming have appeared thanks to wide technological potentialities from combining melt blowing with other techniques for fibrous materials processing. First of all, these are boucle products with a wavy surface resembling fine karakul. They are made by processing melt-blown materials using an original technique uniting needle punching and fiber contraction [1]. Downy ware consists of a fabric formed by multicomponent fibers adhesively bonded at contact sites and fleecy in free portions. The mass of fleecy fibers reaches 60% of that of the fabric. Wrinkly materials are produced by needle punching a layer of fibers that curl and shrink when heated laid on a substrate which is followed by thermal treatment of the sewn cloth. Depending on the regimes of thermal treatment, the needed degree of crumpling and unevenness of the outer layer is obtained. Various ornaments can be made on decorative melt-blown products by bonding fibers at local sites. Highaesthetic quality is imparted to melt-blown textile products by special processing on a carding machine. Original technologies have been developed to confer on fibrous fabrics the texture of materials manufactured by weaving and knitting.

Elastic products are produced from usual isotropic melt-blown materials [4] using special technologies [5] from which they acquire the ability of considerable elastic deformation in two directions, as well as the density, porosity, strength and thickness, elasticity and softness, and visual and palpable qualities approaching those of textile materials.

Miscellaneous textile articles are in growing use today. Melt-blown components are included in the structure of multilayer washable protective sheets. Materials for ladies' underwear on a PA, polyacrylate, and PU base have multitudinous adhesive bonds between fibers due to special treatment. They combine of such properties as resistance to attrition, scratching, deterioration, and tensile strength with extreme softness and elasticity. Electrically heated textile articles present a fibrous mass inside of which wires are installed at a safe distance one from another (i.e., eliminating shorting) connected to a power supply.

Leather substitutes are polymer materials used instead of natural leather in manufacturing footwear, clothing, haberdashery, and engineering products. Conventionally, these materials are composites based on fabrics provided with coatings on a PU or PVC base. Their main disadvantages, low wear resistance and air permeability, were successfully overcome when melt-blown materials appeared.

A novel leatherette that have a density gradient across its thickness is obtained by impregnating needle-punched melt-blown sheets with curable natural rubber or modified by N-methylolacrylamide latexes. To increase its strength nearer to that of natural leather, it undergoes impact loading under temperatures up to the fiber softening point.

Special impregnation procedures for a melt-blown base upgrade the appearance of leather substitutes, for example, eliminate the "orange peel" effect on materials imitating calf leather or add to them a nacreous tint.

The penetrability of leather alternatives should ensure adsorption of perspiration by the footwear or clothes to provide for comfortable wear. To control the penetrability, the fibrous skeleton is impregnated with compositions of polyethylene oxide or components containing NH-groups.

The melt blowing method is highly efficient in molding substrates for leatherette. They can be made as sheets with transversely disposed PU fibers cohesively bonded at contact points (upholstery, car seats) or formed by bunches of adhesively bonded extremely thin fibers (imparting rigidity in bending and shape stability).

Carpets and protective fabrics comprise a backing that ensures toughness and a soft fibrous coating fixed on it.

The backing even of minimum thickness should be rather tough and display dimensional stability. The melt-blown backing makes it possible to use carpeting outdoors and on wet ground [6]. The backing often includes heat insulating layers fixed to the melt-blown fabric by either needle punching or gluing. Original technologies are available for reducing punching force and reinforcing the fibrous backing periphery [1].

Melt-blown components confer a number of specific properties on carpets. Antistatic carpets include a melt-blown backing with an insignificant tendency to generate or accumulate static electricity. The carpet contains up to 30% of carbon-filled fibers that have about 10^{-4}ohm·cm resistivity. In addition, carpets can be stitched with conducting threads. Heat insulating carpets are made of fibrous material fused to a thermally shrinking substrate. On shrinkage, the fibrous upper side of the carpet acquires a greater amount of fibers per unit area. The substrate is often made of a material containing foaming agents that react fusion and shrinkage. Melt-blown fabric can also be modified by foam plastic granules. A great deal of attention has been focused on the fire and explosion safety of textile materials. To reduce the rate of flame propagation, a melt-blown substrate of carpets that looks like coarse linen is needle-punched to metal foil components that can dissipate heat.

11.2 Industry

An extensive range of melt-blown products is used in present-day industry. The major categories of these products are discussed here.

Filtering materials are intended for purifying liquid and gaseous technological and natural media, as well as industrial waste [7]. Based on the dominant filtration mechanism, they can be subdivided into electret, magnetic, adsorption, bactericidal, microorganism carriers, and other materials. The most significant notions of the theory of filtration as well as the range of filtering materials produced by melt blowing have been treated in Chaps. 7–10.

Electric, sound, and heat insulation problems are successfully solved today in many branches of industry by employing melt-blown materials and articles.

The electrical insulation for some industrial equipment can be achieved only by using rigid and sufficiently strong dielectric plates. For this purpose, traditional insulating materials (lacquer, epoxy resin, mica, etc.) are coated on substrates of glass fabric, fiber plastic, or special paper. Melt-blown materials constitute a new means of solving this problem. Insulating plates are formed from aramide fibers (aramide is aromatic PA whose molecules contain amine and carboxylic groups bonded to an aromatic ring; it decomposes at 700°C without melting). A mixture of aramide and other polyamide melt-blown fibers is converted into a pile upon treatment in a carding machine and is then processed into insulating plates by hot calendering. The plates display stable dielectric and mechanical parameters up to 180°C, negligible shrinkage, elevated resistance to bending, and sufficient flatness.

Soundproofing with melt-blown materials presumes the formation of sandwich structures. A facing which includes a melt-blown layer laminated by a neoprene composition filled with fine-ground particles of iron and barium oxides and sulfides reduces the level of acoustic noise in buildings considerably. Such facing fixed on walls or panels outdoes all other traditional building sound-absorbing materials, namely, sound absorbing indoors (e.g., foam plastic, mineral wool) and soundproofing materials for protecting rooms from sound penetration from the outside (sawdust boards, glass wool, porous rubber).

Heat insulation based on melt-blown components solves the problem of protecting engineering and other objects from undesirable heat exchange with the environment.

It is common practice to use melt-blown fabric consisting of microfibers ($d_\mathrm{f} \sim 10$ μm) that have high heat insulating characteristics for manufacturing warmed clothing. The fabric is subjected to pressure treatment during which its density exceeds the initial density and surpasses by nearly fivefold that of other types of fibrous insulation. A mass of heat insulating components is not essential in producing small-size articles, such as gloves or footwear, but becomes a matter of concern for coats, winter sportswear, or sleeping bags [4]. To decrease the mass of the clothes, heat insulating components are made of a mixture of thick and microfibers (to 10 mass%). The former are incoherent and undergo a special coiling action to separate the microfibers in the mixture. Blend fabrics are highly elastic, and their volume and heat insulating parameters are, all other conditions being equal, much higher than those of fabrics consisting only of microfibers.

Separators of chemical current sources (CCS) are capillary-porous partitions installed between electrodes and contain electrode-wetting electrolyte in their pores. Such separators used in CCS make them compact, simplify their production and operation, and hinder mass transfer.

The melt blowing process is convenient for separator production because fibrous sheets are molded with controlled porosity, rigidity, and electric resistance. Of importance is the fact that the process proceeds at the boundary of the thermal oxidation destruction of the polymer which confers fibers surface activity, i.e. the ability to retain electrolyte in the separator pores. The same problem can be solved by electret properties imparting to the separator material [8]. During separator production, melt-blown sheets consisting of PP fibers ($d_\mathrm{f} \sim 0.3$ mm) are premolded under $p \sim 10$ kPa and $T = 140\text{--}180°$C. To add tensile strength to separators and simultaneously resistance to scoring they are formed as mats consisting of adhesively connected sheets.

Production of paper and binding materials is a specific branch of industry where a mixture of a water suspension of plant fibers and target additions (paints, glue, and so on) of the paper mass are processed. Melt-blown materials here play the role of both the production means and the raw material.

Meshes of paper-making machines simultaneously support the wet cloth of the paper mass and container the water which runs off and is removed by pressing, evacuation, and drying. The melt blowing technique makes it possible to produce unusual meshes where the paper contacting surface layer is formed from flat fibers [7]. Such fibers are rectangular in cross section and their sides are in a ratio of 3:1 and more. This increases the contact area and the mesh to cloth adhesion, enhances paper cloth pressing, and speeds up its dehydration. The upper layer of the mesh is commonly fixed on the fiber bearing layer by needle punching.

Materials for binding need to be wear resistant to protect a book binding from crumbling and paint from tarnishing. Therefore, special technologies for processing melt-blown blanks of binding material have been developed. According to one of them [1], a fibrous sheet of a carefully controlled basic weight and density is formed by melt blowing and is afterward subjected to calendering under optimum regimes. The sheet develops a considerable resistance to delamination, abrasive action, impermeability to light, and improved printing ink adhesion.

Another method includes impregnating melt-blown blank with a binder capable of thermally induced polymerization. Upon binder coagulation, a water repelling agent is applied onto the blank as a paste or foam which is then dried or hot pressed. Paint on such surfaces intrudes between the fibers and confers elevated wear resistance on the binding.

Conveyor belts, drive belts, hoses, tracks of vehicles, etc. usually achieve their structure from the fibrous reinforcing element found within the elastomeric binder (rubber, PU, PVC, etc.). Reinforcements and elastomer are bonded to each other. Melt-blown materials used as reinforcements add negligible specific mass to the articles but impart considerable strength exceeding that of alternatives, plus elasticity, and resistance to dynamic tension, compression, and creep.

11.3 Construction

A wide group of melt-blown materials is used, among other building materials, in the construction and maintenance of buildings and structures.

Floor materials differ in their structure depending on the design of buildings.

Asphalt floors in industrial areas are made by gluing asphalt-modified fibrous plates on the floor backing surface. Melt-blown sheets mainly of PP fibers are saturated with molten asphalt (softening point $\sim 90°$C). Evenly saturated 4–5 cm thick sheets are obtained by dipping them in a certain manner into an asphalt bath with a temperature gradient ($140°$C on the melt surface and $150°$C on the bottom) where molten asphalt squeezes air from between the fiber spaces. Fibrous materials with asphalt coatings are formed by short-term dipping of melt-blown sheet blanks ($h = 4$–8 mm) into the bath and further removal of excess asphalt by a rake so that an asphalt layer 0.5–1.0 mm thick is left on both sides of the sheet. The bottom side of such sheets is coated in addition by a catalytically active asphalt composition that provides good adhesion to the floor. The upper side is often covered by rubberized asphalt for waterproofing. When a layer of foam PU ($h = 10$–15 mm) is applied on the upper surface of asphalt-modified sheets, the materials acquire heat-proofing properties. They can be dressed up with decorative coatings, e.g., on a PVC base.

Floors for gym-halls must meet a unique complex of requirements; they must have a rough surface and create a feeling of comfort when walking barefooted, be dimensionally stable and show low residual deformation under compression, provide for ball rebound similar to a wooden floor, possess frictional characteristics to allow safe running and stopping, self-restore upon being imprinted by spikes, and not become slippery when wet.

The material for such a floor has a multilayer structure; every layer fulfills a certain function, and the combination of layers provides for realization of the set of characteristics named. Its fibrous melt-blown substrate is made of PA, PP, polyester, polyacrylate, or mixtures. The substrate is bonded with a melt-blown cloth by needle punching so that the substrate fibers penetrate through the cloth and extend 5–6 mm above the other side of the cloth. The fibers are impregnated with latex and bonded by curing with a elastomer PU layer upon which a PU-based outer coating is applied.

Decorative floors are elastic because a melt-blown cloth is used as a substrate. Cloths are placed on both sides of a reinforcing glass-fiber sheet, and the three-layer mat is fixed by needle punching. A latex-based coating is formed on the upper surface of the mat, which can be then decorated by an embossed pattern using a foam plastisol.

Heat- and soundproofing coatings as well as facing and wallpaper on a melt-blown base are employed in buildings where they simultaneously have technical and decorative functions.

Relief fibrous facings are made on walls from needle-punched melt-blown cloth and linens Decorative pictures are embossed on the fibrous mat thus obtained using heated rollers. Fibers are heated to their melting point and then premolded at local sites to obtain an ornament or picture.

Sound-insulating wallpapers are also multilayer products whose face is formed from a melt-blown sheet based on antipyrene-modified PP or polyester and is fitted with decorative trimming. A microporous layer of a synthetic foamed resin is secured to such a sheet, and a melt-blown cloth formed from fibers modified by a pressure-sensitive adhesive composition contacts the wall layer. The elements of this wallpaper are bonded using ultrasonic welding or are fused together by passing current through to generate the effect of a quilted article.

A fibrous coating pasted to a wall contains a melt-blown mat, including a pair of fleecy cellulose cloths, similar to those used in throwaway washing napkins, and a reinforcing mesh between the cloths. The mat adheres to a technical paper sheet over whose surface a decorative pigment layer is spread. The sheet is impregnated with a synthetic resin that fixes the pigment particles and fleecy cloths in mesh holes. This system is laid on the wall and rolled down for the binder to adhere to the wall, after which the technical sheet is peeled off.

Supporting and waterproof mats are used in construction as different purpose structural elements.

Load-carrying mats are produced by depositing of polymer-melt filaments emanating from a multihole head on a liquid surface. The liquid performs the functions of precipitating substrate and cooling agent. The fibrous mass plunges into the liquid that passes between guides placed under the liquid surface, thanks to which a given density and texture of the fibrous mass are attained. Following the technology described mats of unlimited length and controlled thickness or density can be formed in response to the required flexibility and elasticity of structural elements. The mats are employed as light panels for slight load-bearing walls, insulating dividing walls, overhead ceilings, or exhibition wainscots.

Waterproof mats consist of a melt-blown framework impregnated with sulfur and a bitumen-based composition. Liquid bitumen is modified at 120–160°C by sulfur (sulfur to bitumen ratio is 15:85 to 30:70 mass%) or, if necessary, by solvents, pigments, herbicides, fungicides, or other agents. The fibrous framework, either dry or moistened (upon treatment by an adhesive), is impregnated by the mass described in a viscous-flow state. Articles thus produced are additionally treated by latexes and undergo calendering after which they can be successfully used in construction as light structural members of systems for water- and moistureproofing ceilings, walls, and floors.

Linning materials for reservoirs prevent liquid from soaking through to the load-carrying walls. The problem of protecting walls becomes of paramount importance in the storage of hostile media. The manufacture of reservoirs made entirely from inert materials can hardly be justified from the viewpoint of economy. It is worthwhile to make them from conventional structural materials and to insulate against such liquids by a resistant lining.

Reservoirs where liquids are stored layer by layer where the upper is hostile and the lower inert are lined in the upper part by a belt of chemically resistant material. The belt is made of melt-blown sheets saturated with binders indestructible in hostile media, namely, nitrile rubber, PU, and acrylonitrile-butadiene-styrene. The lower part is faced with melt-blown sheets saturated with asphalt.

A polymer laminate is used to line reservoirs for storing hydrocarbon liquids. It involves a PP melt-blown sheet (specific mass 100–250 g/m^2, produced by method [9]) welded to a 0.2–1.3 mm thick PE film. To raise the strength of the lining, fibers from the reverse side are fused together. The following laminate modifications with fibers fused from both sides can be used: film + melt-blown sheet; film + sheet + film; sheet + film + sheet.

11.4 Medicine

Medicinal materials and products on a melt-blown base include surgical and dressing means as well as medical equipment.

Surgical materials are employed in operations involving tissue dissection.

Surgical serviettes are two-layer fibrous structures whose lower layer formed by a melt spinning procedure consists of orientation-reinforced fibers 12–25 (to 50) μm in diameter. The upper layer is produced at a small distance L by melt blowing (see Fig. 2.1) and includes short microfibers 2–6 (to 10) μm in diameter. The layers can be molded from polyolefins, polyamides, polyesters, PU, and other thermoelastoplasts. The addition of pigments into the extruded material makes it possible to obtain colored layers. The relation of the upper and lower layer masses varies within 0.2:1 to 4:1, depending on the strength of the napkin needed. Layers put together undergo calendering and are passed between a pair of heated rollers, one of which possesses a relief as regularly spaced projections. During rolling, some discrete areas of the adhesive bonds between the fibers of the two layers are formed in the zones of the projections. They occupy about 10–30% (up to 50%) of the whole area of the cloth. From a technological standpoint, it is advisable to maintain the softening temperature difference in the upper (T_1) and lower (T_2) layers equal to $T_1 - T_2 = 10\text{--}40°\mathrm{C}$.

Five-layer surgical towels are disposables that have high absorptive capacity; they absorb moisture perfectly but do not admit it through the layers. Usually, about 15–30 g of water remains on surgeon's hands after washing before an operation. The towel of 30 g mass absorbs 4 to 7 g of water per 1 g of its mass.

The outer layers of a surgical towel are laminated and include melt-blown PP cloths of an average density (specific mass ~ 15 g/m^2) glued by a latex to a paper napkin. The napkins are treated with special compositions to improve their wettability and resistance to abrasion. The outer layers acquire an embossing which perfects the surface texture and makes the product softer. The middle layer (~ 30 g/m^2) of the towel consists of ultrathin melt-blown PP fibers. The layers are bonded to each other by the edges, lengthwise and widthwise, using seam resistance spot welding. The regular position of welding spots improves the product appearance.

Such a towel meets practically all of a surgeon's requirements; it is much cheaper and softer than the cotton one, pleasant to the touch, dries perfectly, is resistant to abrasion, and is strong.

Vessel endoprostheses are tubular elements implanted into the blood circulation system by surgical methods. They must meet contradictory demands, i.e., on one hand, they must be tight but allow for blood circulation without leakage and, on the other hand, they should be porous to create conditions for soft tissue growth into the tube walls to form a single vascular system.

The problem is solved by formation of initially tight tubular elements whose walls are furnished with holes shut off by plugs made of a special coagulant, whose composition is the *know-how* in many patents at the base of artificial vessel technologies. Upon endoprosthesis implantation, the plugs resorb, and soft tissues start to grow into the tube walls. These coagulants in the wall structure of artificial vessels are, however, unsafe because of thrombogenesis. To reduce the danger, the plugs are made of a material that can retard blood coagulation and adhere well to the coagulated fragments. Such precautions reduce the probability of thrombus escape and travel through the blood system.

Vessel endoprostheses on a melt-blown base are produced by winding a fibrous blank onto a conical mandrel followed by hardening and needle-punching it. A continuous technological process has been developed to produce tubes with a homogeneous structure 4–5 mm in diameter and 0.5 mm thick. Generation of a negative electret charge on the inner side of man-made vessels is known [10] to reduce the risk of thrombogenesis. The use of electret melt-blown components in endoprostheses helps to attain this aim with minimum expense.

Bandages and splints are medical materials designed to immobilize injured or sick internal organs, bones, joints, or soft tissues during patient transport and cure.

Nonwoven spun bandages based on melt-blown (or spun-bond) materials have now superseded fixed gauze dressings. The latter are insufficiently strong, their absorptive capacity is rather low, and they fray quickly. A melt-blown bandage displays higher values of air permeability and strength, along with elasticity. With careful handling such bandages absorb blood like a sponge, do not practically wear out, and can be reused after washing and drying. The optimum density of a standard thickness bandage is 20–40 g/m^2; at lower density they are insufficiently strong, and at higher density are strong in excess. By optimizing substrate displacement during melt blowing, most convenient anisotropic bandages are formed whose lengthwise and widthwise elasticity can be controlled.

Water-curable splints look like a fibrous cloth spun onto an immobilized part of the body which solidifies when wet. The fibrous base of the splint consists of two layers where the woven cloth is doubled by a melt-blown cloth. The latter is impregnated with a paste which includes a mixture of ion-salt inorganic material (e.g., aluminosilicate glass powder) and a dispersion or solution of a polycarboxylic acid in an organic solvent (methylethyl ketone, cyclohexanone, methylene chloride). The substrate impregnated by the paste is then dried by evaporating the solvent. The amount of modifier precipitated on the fibers is 60–90% of the splint mass. The paste can also include a binder which increases the modifier adhesion to PVC or PVA fibers (0.1–1.0% of the paste mass). The penetrability of the fibrous melt-blown base by the paste simplifies the impregnation process, and the splint thus obtained becomes airpermeable, which is valuable for the accelerated healing of injuries. The splints are simple to use, rather strong, solidify upon wetting at room temperature, are resistant to hot and cold water, and are transparent to X rays.

Materials for curing burns are suitable for solving the main problem of burn wounds. Skin on such wounds is absent or damaged so that its major physiological functions are not carried out. It can not be a barrier to infection or prevent liquid and protein loss. The evaporation rate of water through sound skin is 2 $mg/(h{\cdot}cm^2)$, whereas burnt skin loses 25–45 $mg/(h{\cdot}cm^2)$ of water, and this figure is even higher during the first hours after the trauma.

Membranes of silicone rubber are only 50–60 μm thick and are fit with a special surface texture on either one or both sides. They are fabricated by applying a silicone rubber coating to a fibrous substrate made of a freely soluble polymer. When the rubber cures, the substrate is removed by dissolving the fiber in certain solvents. According to a number of parameters, this biologically inert membrane is an analogue of human skin. Its elongation is above 100% in all directions, its porosity reaches 60%, and its water evaporation rate is 2–20 $mg/(h{\cdot}cm^2)$. The texture of the membrane surface layer features numerous microridges formed by rubber leaking into cavities between substrate fibers and microvalleys formed after fiber dissolution. The texture described provides perfect adhesion between the membrane and the wound and imparts to the membrane the property of absorbing foreign particles in the wound. These membranes are convenient, can be transparent or semitransparent and endure prolonged storage.

A nonwoven protein material is produced by melt spinning (see Fig. 3.23) water solutions of a collagen purified from peptides (organic substances based on aminoacid residues connected by peptide links are formed by enzymatic splitting of proteins). Fibers are cut into short chips from which a cloth is formed of 0.05–0.4 g/cm^3 density by either a papermaking machine ("wet" technology) or using a collagen binder ("dry" technology). The cloth is then treated by a tannin to increase its water resistance, is cut into sheets, and sterilized. Such a material cleans a wound by interacting with the organism's fluids and forms a protecting crust. No other man-made materials can exert such a unique effect on burn wounds. This collagen material demonstrates adsorptive characteristics and permeability to water vapors similar to human skin. The collagen dressing can be left on the wound because it is self-resorbing.

Napkins for burnt wounds are laid between wounds, and a sheet or blanket does not let them stick to the wound. They are also spread on surgical tables and other medical equipment that supports burned parts of a body during wound cleansing. The napkins include three layers of laminated material whose outer layers are antiadhesive; the middle one is elastic because it is made of polyurethane open cell foam. Antiadhesive layers are formed as cloths consisting of viscose fibers oriented longitudinally and fixed by acrylic resin applied on the outer surfaces of the cloths. So, the cloths obtain a smooth outer surface and a rough reverse side that is glued to the middle elastic layer. To avoid a shift of separate fibers, they are fixed on smooth surfaces by the resin, and specific properties of the viscose make the contact between the napkin and the wound painless. An original lamination technique by heated rollers to produce napkins employs melt-blown components.

11.5 Packing

Packing materials and articles are numerous and their range is commensurable with that of industrial and agricultural products that are packaged. Due to their unusual properties, melt-blown materials have occupied a separate and deserved place in this sphere which is constantly widening because traditional packages for liquid and bulk cargoes (cans, boxes, and rigid containers) are easily damaged, especially at limiting pressures experienced by the contents or when the cargo volume is large. In addition, transportation of these empty packages is highly inefficient. Melt-blown materials have made it possible to improve packing reliability, reduce its weight, and return it to a producer collapsed as a stack of fibrous sheet materials.

Materials for marine containers used to ship bulk cargoes (soybean, nuts, granulated industrial goods, etc.) consist of four adhesively bonded sheets based on isotactic PP fibers made by process [11]. The mass of a sheet constitutes about 22–28% of the total material mass. The sheets also include a thermoplastic melt-blown binder. PP fibers in the inner sheets are mutually perpendicular and are located at a 45° angle to the fiber direction in the outer sheets. The sheets are glued together by rolling with hot-vapor heating. Binder spreading in the sheets depends on their position relative to the vapor source upon rolling. Bags made of such materials endure falling from a few meters of height, do not laminate, and possess low specific mass and porosity which prevent bulk cargo losses.

Containers and bags for medical instruments are made of either melt-blown or spun-bonded polyolefin cloths. It is convenient when one of the containers walls is made of a transparent plastic, and others of nontransparent elastic melt-blown material. The fibrous and transparent sheets are bonded together, and the permeability of the fibrous walls is controlled by the combined action of heat and pressure. Under optimum hot premolding regimes, the needed strength of containers is attained along with their ability to "breathe" that is necessary for storing a number of goods.

The material for tight containers is a multilayer sheet consisting of the following adhering to the other main layers: a melt-blown or spun-bonded outer fibrous layer, made of a foil middle layer, and a inner layer based on polyolefin film. An outer layer based on polyolefin or polyester displays the highest static and fatigue strength. Its thickness and specific mass (60–80 g/m^2) are selected according to the mass and volume of goods to be packed. A layer of foil 6–25 μm thick makes the container resistant to UV radiation and impermeable to gases and liquids. An inner layer of film 40–140 μm thick is welded for tightness after filling.

Some modifications of the material employ additional adhesive layers, e.g., outer layer fusion with the foil by a PE film or outer fibrous layer protection from scratching by gluing a kraft paper, etc. Upon filling, the container may be evacuated or filled with inert gas and sealed by welding an inner layer.

Antistatic frictional coatings are applied to packages that have a melt-blown or some other fibrous outer layer on a polyolefin base by spraying a composition consisting of an aqueous wax dispersion (up to 10%), surfactant, antistatic agent, and water-insoluble resin which wets polyolefin. When the coating dries, the fibrous package acquires an antistatic quality and water repellency. Such coatings prevent slippage of containers filled with goods when stockpiled. A steel plate (mass, 2.3 kg; contact area, 106 cm^2) packed in a melt-blown PE net covered with an antifrictional coating has a sliding angle over an inclined stainless steel plane of about 30°. It is possible to apply pigments on the coating to produce a colored package.

Foodstuff packages based on melt-blown materials solve problem of transporting meat from the slaughterhouse to the processor and distributor of meat products. Meat cut into slices usually contains sharp projecting bones, which could damage a package. This is undesirable. The package should be tight to eliminate meat drying or getting soiled and to prevent insect access.

The problem is solved by a special packaging material involving an outer PA film, a melt-blown cloth, and a meat contacting film based on polyesters, or ethylene and vinyl acetate copolymers. The layers are joined together by an adhesive, e.g., wax, which saturates the melt-blown layer. The material becomes resistant to cutting, rupture, or scratching. The inner layer fuses readily, thus sealing the meat package.

11.6 Protection of Products and Environment

The title defines the sphere of application of the last group of melt-blown materials.

Cleaning and polishing materials are intended for cleaning, selective washing, and polishing articles in the industrial, construction, and household domains.

Napkins for furniture polishing are disposables made of melt-blown cloths impregnated with a polishing composition. The cloth of ~ 40 g/m^2 density absorbs up to 600 mass% of liquid. The content of the polishing composition constitutes about of the half absorbing capacity of the cloth (ratio of composition–cloth mass is from 1:1 to 3:1). The composition is an aqueous emulsion of mineral oil (4–65 mass%), including also a silicone liquid and emulsifier. Wiping with napkins imparts a gloss to furniture and smoothes the lacquer coating. The napkins are resistant to abrasion.

A wiping material that contains surfactants is used for wet cleaning and for washing machines and equipment, as well as buildings. It is based on a melt-blown cloth pretreated with water (e.g., splashed) containing surfactants and is rolled by rollers with a clearance less than the cloth thickness. Rolling brings about a mutual shift of fibers that forces surfactant to spread evenly across the cloth volume. Upon drying, the wiping material is stored without any change in properties for a prolonged time. Before usage, it is moistened slightly to facilitate dust removal from machine parts.

Napkins with a rubber foam coating to clean windows and glass facings consist of a nonwoven fibrous sheet based on a melt-blown and needle-punched substrate fitted with a two-layer coating on a butadiene-nitrile rubber base. The sheet is formed from a mixture of fibers (1:1) that can absorb moisture (cotton, viscose) and synthetic thermoplastics (polyesters, polyolefins, polyamides, and so on). The sheet surface is furnished with a discrete coating based on a latex that is applied by an offset duplicator. The coating is made as a pattern of circles or squares, cells, twisting lines, etc. projecting across the surface and occupies up to 70% of the sheet area. When glass is rubbed by a moistened napkin, the rubber projections of the napkin absorb water and dust and add wear resistance to the napkin. The fibrous base of the napkin absorbs water, and dust accumulates within the depressions between projections. The mass of the napkin is about 200 g/m^2, and the share of 30 g/m^2 on each side belongs to the rubber foam coating.

Sheets that selectively absorb certain components of liquid blends are designed mainly to isolate hydrocarbons (petroleum, oils) from aqueous mixtures. They include a pair of layers from nonwoven materials chiefly on a cotton or viscose base and a separating and strengthening interlayer of melt-blown polyolefins. Cotton layers are modified by oil-absorbing agents (wax, asphalt, rosin, fatty acid anhydrides, tall oil derivatives, etc.), and their surface is treated by water repellents, including silanes, olefins, and ammonia complexes. The layers are glued by an adhesive (starch, PVA, proteins, latexes, resin solutions, etc.) and cut into 30×30 cm squares after setting. Material thus obtained is rather flexible, strong, and lightweight.

Mats for collecting petroleum look like fibrous sheet materials with a cellular structure. A highly dense cloth contains projections (either spherical or elongated) whose density is minimum while the free volume is maximum. The materials are produced by depositing melt-blown fibers on substrates with a special forming surface where depressions with $D = 5$–10 mm and more are at a regular from one another. The dense cloth adds strength to the mat (by reinforcing it with wire or glass fibers) in contrast to padprojections that have a stable shape and a friable multilayer structure that can absorb petroleum products. The spray distance and projection diameter are related as $L/D = 5{:}1$–$10{:}1$ (commonly $L > 4$ cm). The spray gas pressure is $p = 0.1$–0.2 MPa, fiber diameter $d_f = 1$–100 μm and length $l \sim 10$ cm. The mats are often of a sandwich type whose outer layers are strong enough to protect the middle adsorptive one from mechanical damage. They are formed by multihole spray heads [12, 13], that have three rows of spinnerets of which the middle row has the smallest diameter and hole pitch.

Materials hindering soil erosion are intended to prevent water- and wind-induced disintegration of the upper soil layer.

Melt-blown mats for keeping soil on steep uncovered slopes, particularly, along highways and canals, are composite products based on wire or plastic net reinforcements. The reinforcement is covered by a melt-blown layer so that a portion of the fibers protrudes into the net cells and adheres to them. The fibrous coating is deposited on both sides of the net. Such mats are installed on a damaged slope and are powdered with soil. Stony slopes are covered by soil with the addition of grass or seeds and a fertilizer. Such measures lead to fast landscaping of bare slopes and make them resistant to soil erosion.

Camouflage materials are aimed at changing the external view and misrepresenting the outlines of disguised objects.

Camouflage meshes are based on a reinforcing cellular backing on which various shape and color spots and strips are deposited by the melt blowing technique.

Radio-absorbing meshes hamper radio detection and the range of military and industrial objects. The mesh consists of a base made of strands of PA or polyester fibers and conducting fibers (2–10 mass%) made from either stainless steel or graphite. The strands are interlaced into parallelogram shaped cells with 100–105° and 75–80° angles. The base is covered on both sides by melt-blown polymer coatings, which are fastened mechanically in the cells. The mesh acquires the form of a flexible cloth whose radio-absorbing characteristics are not impaired by bending or folding.

Means for protection against biological injury of industrial equipment and agricultural products consist of components in the form of specially designed melt-blown materials.

A material to protect wood against spoiling by marine parasites is made as a melt-blown sheet on a PP base with a pore size from 150 to 180 µm. Wood placed in a bag made from such material is not spoiled by worms even after long exposure in seawater containing these parasites in abundance. When used wood populated by parasites is placed in this package, the worms soon loose their activity and die. Tests carried out in one of Alaska's gulfs populated by teredo worms have proved the high efficiency of melt-blown material in guarding pine wood [1].

An insecticidal material is produced as a melt-blown cloth (85–97% porosity, 0.5–5.0 cm thick) impregnated with special additions. The cloth is made of structural thermoplastics, including polyesters (chiefly PET), polyamides, PVC, and polyacrylates. Its surface is saturated with volatile insecticides: pyrethroids, chlorinated hydrocarbons, or naphthenates, and others. The concentration of insecticides in the material varies depending on their efficiency and the minimum level of toxic effect on insects during a given time. When the cloth is 1 cm thick, the typical concentration varies within 1–20 mg/cm^2 of the material. The cloth can be modified by an attractant, a substance with a specific smell for attracting insects.

The material is left in a place inhabited by certain insects for some time. The cloth is compressible and possesses memory, so it can be placed in narrow gaps, e.g., behind kitchen devices not blocking the gap but making a lethal concentration of insecticides for insects. As soon as the problem with insects is solved, the cloth is removed and thrown away, and the remainders of the insecticides left by the cloth continue acting for some time.

This review of the fields of application of melt-blown materials and products on their base does not, certainly, embrace all of the applied aspects of the melt blowing technique. It suffices to recall the use of melt-blown adhesive linings in light and textile industries, deicer systems for planes and helicopters, nonerasable inscriptions, labels, and many other melt-blown elements of modern clothes. There is no doubt that melt-blown materials have come into our lives for years in the form of numerous technical products of contemporary society and they are unlikely to be superseded for the present by any other materials or articles.

12. Ecological and Social Problems

By now, the volume of consumption of melt-blown materials has reached a critical point in response to the perceptible influence of those products on society and the state of the environment.

However, the public recognition of successes in this field is not as prominent as it could be because a major part of the melt-blown articles used generates a heap of wastes and solid rubbish whose use has grown into an acute global problem. Regeneration and recycling of melt-blown products constitute part of a problem of polymer waste recovery. It is complicated by such specific features as the nonequilibrium state of fibers and their intense interaction with other components and the environment.

12.1 Solution of Ecological Problems

Ecology is the science of studying vegetable and animal organisms, formed by their communities and their relations with the environment. Unfavorable ecological conditions on the earth in the second half of the twentieth century have put the problem of protecting nature in the forefront of human activities and have forced reevaluation of the final targets of human economic activities within the framework of sensible and reasonable use of natural resources. Some time ago, the Russian scientist V.I. Vernadsky considered progress in nature in interrelation with the human history by distinguishing the anthropological factor of natural variation (i.e., based on the notion of "humans" as a conceptual category). He put forward an idea that the biosphere is developing in dialectic integrity with natural and man-made factors [1]. In the twentieth century, ecology has formed a scientific base for rationally treating nature and safeguarding living organisms. It split into some branches, including *engineering ecology*, a science studying the reasonable limits of human activities. Melt-blown materials turned out to be a highly efficient means of solving a number of questions posed by engineering ecology. Some aspects of their use in cleaning industrial gas emissions and wastewater, in draining swampy lands, and in picking up spilled hazardous liquids are discussed here.

12.1.1 Purification of Industrial Gases

Purification from the numerous contaminants found in the air basin of towns and in the countryside is an objective necessity of modern industrialized society [2].

Since the beginning of the twentieth century to today the emission of harmful substances into the atmosphere from all power stations on the earth has grown at least by an order. The growth of the main components of these emissions can be characterized by the following numbers (in million tons/year): dust-like matter, from 10 to above 100, sulfur dioxide, 12 to 150, carbon monoxides, 30 to 300, and nitrogen oxides, 1 to 50 [3]. Contamination of the atmosphere by industrial wastes has brought the danger of injuring the ozone layer – a specific protecting jacket which guards our flora and fauna from the detrimental action of cosmic rays. Air pollution could bring about climate warming and will change the ecological balance of the earth significantly as ice melt and the level of the world ocean rises.

The filtration of gas emissions is a multistage process involving the capture of solid and liquid contaminants and the removal of ecologically unsafe gaseous impurities from emissions by physicochemical methods (absorption, adsorption, and chemisorption). Filters for cleaning gases from solid particles of a typical structure usually include three components: (1) a preliminary cleaning filter that traps coarse dirt particles and possesses high mud capacity; (2) a main filter that captures particles 0.5–5.0 µm in size which passed through the preliminary filter and are very hazardous to the facades of houses, presses and precision machine parts; (3) a filter for gases cleaning from submicron size particles where mostly electret materials are employed [4]. Each of the components named incorporates melt-blown materials. To absorb hazardous gases from industrial emissions, it is worthwhile to use the adsorptive melt-blown materials described in Chap. 9. Among the strictest limitations for melt-blown PFM in gas purification systems is the rather low upper service temperature of 90°C for PP and 150°C for polyesters.

There are, nevertheless, numerous examples in international practice of highly efficient operation of melt-blown materials in industrial gas cleaning systems. The Verseidag Company in Germany has developed a unified design of a gas filter successfully operating in environmental protection systems. Original modules for superfine gas cleaning systems were created at Vokes (GB), Ultrafilter (Germany), and Filtrair (Netherlands). Widely known in the field, 3M (GB) and Lydall (USA) employ carbon-modified adsorptive melt-blown materials for isolating ecologically unsafe components from gas emissions. The excellent reputations of these companies speak in favor of melt-blown PFM as efficient components of industrial gas cleaning systems.

12.1.2 Wastewater Purification

The purification of industrial, domestic, and rain sewage containing hazardous contaminants is an urgent problem of modern nature management.

The main source of soil contamination is considered to be industrial wastes because the product yield of the technological chain "raw material — target" rarely exceeds 10% but usually corresponds to merely 1–3%. From the viewpoint of ecology, most dangerous wastes are industrial effluents. Soil is gradually contaminated by deep water spilling into drilling wells (salinization, swamping), by the effluents of chemical plants (acid, alkali, pesticide production, etc.), those of plastics and rubber processing plants, gas and coke-chemical plants, wood processing, textile, and tanning and paper-making factories. The world industry consumes more than 10–12% of fresh water from global circulation for technological needs. In addition, 30% is used for effluent dilution in neutralizing which is, unfortunately, not always and not fully efficient. The intense growth of urban areas has led to contamination of soils by domestic wastes, because about 9 liters of effluents are generated by each citizen each day.

Requirements for cleaning effluents differ considerably from enterprise to enterprise. The common problem for all of them is petroleum admixtures in effluents, which create additional difficulties in separating suspended or sedimentary solid particles found in solutions with a wide pH range. Almost all types of standard equipment for neutralizing effluents (gravitational, and coalescent separators, centrifugal and gravitational precipitators and filters, systems for chemical treatment of effluents) incorporate filtering elements at the stage of pneumatic aeration and at the last stage of effluent cleaning by adsorptive, bactericidal, and microfilters. Pneumatic aerators with melt-blown elements developed in the MPRI are successfully operating in systems for effluent cleaning with the following service characteristics: aeration efficiency 3.6 kg/kWt·h at an air discharge of 3.4 m^3/h and oxygen dissolving rate up to 44 g/h [5]. In Chap. 9, data on the performance of adsorptive and bactericidal melt-blown FM are cited.

A promising direction in wastewater cleaning is biodestruction of admixtures by microorganisms in biofilters where biophysical and biochemical processes simulate water self-cleaning in nature. Chap. 10 deals with the profound potentialities of melt-blown materials as microorganism carriers in a new generation of biofilters. The fields of their application are continuously expanding and have occupied a leading place in engineering ecology.

Equipment for cleaning effluents manufactured at such companies as Osmonics Inc. (USA), Schenk Filterbau (Germany), Sartorius Ltd., Wemco and 3M (GB) also contains FEs made of melt-blown polymer materials.

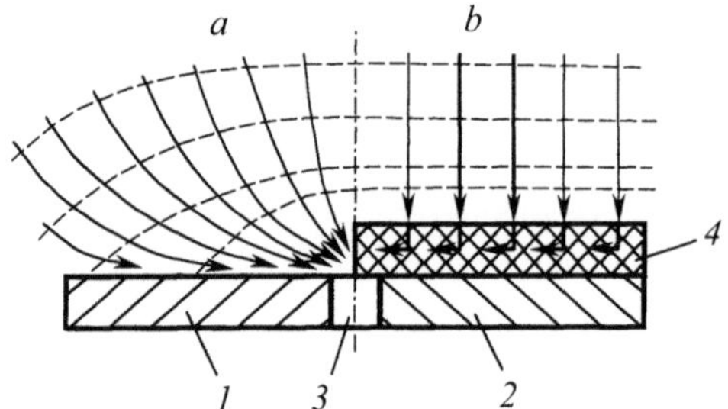

Fig. 12.1. A diagram of subsoil water flow to the drain: (**a**) without filter; (**b**) with a filtering shell; *1* and *2*, pipes; *3*, gap; *4*, filtering shell. Arrows indicate the flow direction; dashed lines are discharge isolines

12.1.3 Melioration

Melioration includes a combination of measures for land reclamation on which water and air regimes are considered adverse. One of the popular types of melioration in the middle belt is drying, i.e., moisture removal from root-inhabited zones of soil. Excess water is removed using drainage systems, the main elements of which are drains, so-called man-made underground channels consisting of either porous ceramic or perforated plastic pipes laid in the soil. To protect the pipes from silting (pipes clogging with silt), they are isolated from the soil by filtering shells.

Another important function of the shell is contribution to the efficiency of drainage systems. As shown in Fig. 12.1, subsoil water streams concentrate near the drain inlet when there is no filter (a) thus leading to increased inlet resistance of the system and hydrostatic pressure loss. The filtering shell directs subsoil streams along the normal to the drain making water discharge more uniform (b) and reducing pressure loss at the drain inlet.

In the 1980s an original design for drains was developed at the MPRI on the basis of plastic pipes with a melt-blown coating. This has improved the productivity of drainage systems by trenchless drainlayers significantly. The performance of the new drains was rather high: filtration rating of melt-blown coatings, 100–200 m^3/day; maximum water runoff over the drain, 0.87 cm^3/s.

12.1.4 Oil and Chemical Sorbents

One of the most difficult problems in engineering ecology is collecting the spills of ecologically unsafe liquids. Oil is necessary everywhere in modern industry either as a raw material or lubricant or as a power source. Unfortunately, danger always exists of spilling during refining, storage, or transporting. Spillage might occur in oil well drilling, pipeline and cistern failure, dumping bilge water from ships, and carrying effluents out into the sea, which threatens all forms of life. One ton of spilled oil is known to form a continuous film 2.6 km^2 in area on water. As for ice, it absorbs oil up to one-quarter of its mass. When the oil concentration in water exceeds 800 mg/m^3, metabolism is suppressed bringing about the death of plankton, the abrupt reduction of oxygen liberation by ocean algae, and worsened fish breeding gain [3]. Burning of spilled oil threatens every live creature on the earth and in the water.

Spills of phenol, semiproducts of oil and wood chemistry, and other chemical branches represent a great danger to the existence of flora and fauna. Being strong antiseptics, phenolic waters intrude into biological processes taking place in water and on land. During recent decades, the danger of contamination with synthetic surfactants widely used in drilling gas and oil wells, minerals dressing (flotation), and the production of detergents, lacquers, paints, pesticides, and food products has risen perceptibly. The surfactants form stable foams that reduce the efficiency of biological cleaning of both soil and water, and stop algae and other vegetative growth. In this connection, spills of hazardous liquids should be treated as serious accidents independently of their dimensions. With increasing scales of spills, the border between the ecological aftereffects of spills and natural disasters that can lay waste to the human habitat vanishes.

Information on the efficiency of melt-blown PFM as sorbents of oil and other hostile media is given in Sect. 9.2. The melt blowing technique serves as an innovation base for a wide range of highly efficient sorbents produced by the 3M company [6].

Oil sorbents of 3M (trademarks T and HP) are highly efficient on land, and such properties inherent to their melt-blown base as water repellency along with perfect wettability by oil-products make them ideal for collecting oil from a water surface. The 3M Company manufactures sorbents of various kinds, including

- bands recommended for collecting small oil spills or other hydrocarbons;
- rolls for spills on great areas both on land and water;
- pads for absorbing oil in hard to reach zones (bilge, tanks, cisterns, gutters, settlers, etc.);
- booms to remove oil films, also for enveloping and absorbing oil spills on inner water routes, in harbors, and ports;
- sweeps, i.e., melt-blown bands which collect oil when pulled over a water surface by a fleet.

Chemical sorbents of 3M (trademark HSRK) are intended to eliminate emergency spills of a wide range of hostile media. They are produced similarly to oil sorbents either as sheets (small spills, final operations on elimination of large spills) or pads (sorption capacity, to 2 L of liquid), minibooms, or rolls.

Analogous materials of grades SXT, UXT, CC and KC are produced by Sorbent Products Co., Inc. (USA and Belgium). Melt-blown meshes and seines of this company can be reused many times by returning the oil removed from water into fuel tanks [7].

Melt-blown oil sorbents make it possible to eliminate spills with a minimum of material and labor consumption; they can be regenerated and are easily burned leaving less than 0.02% of sinter. In use, chemical sorbents acquire the property of absorbed liquids, so they should be handled with precautions like hazardous liquids. They are disposed of according to state regulations.

Data noted earlier are limited only to the main fields of melt-blown material use in solving ecological problems. Beyond these bounds, such urgent aspects of the problem remain as elimination of domestic wastes by disposables, filtration of running water, protection of breathing organs, and some other questions treated partly in previous chapters. Summarizing, it should be emphasized that melt-blown materials in their technical and economic aspects are a highly efficient means of successfully solving many problems in environmental protection.

12.2 Regeneration, Utilization, and Burial

In 1977 in Geneva, 34 European states adopted the Convention "On prohibition of using military or any other hostile means in respect to natural environment." Two years later, another document was signed there: "Declaration on low-waste and waste-free technology, and waste reclamation." In fact, all countries, at to the present moment, have restricted legislatively and in the spirit of these documents, methods of waste storage, control, and reclamation and have fixed responsibility for contamination and the use of dumps and installations for refuse incineration. The time when manufacturers could peacefully return to other activities after collecting their wastes has passed irrevocably. Hereinafter, some aspects will be discussed of the way this ideology has touched upon the production and application of melt-blown materials.

Regeneration is the restoration of the initial properties of a used up product.

Melt-blown filters belong to a group of depth (volume) filtration filters, which are, as a rule, not subjected to regeneration. For partial regeneration, only those filters are suitable which are clogged by contaminants apt to be transferred into a movable phase by special treatment and then removed from the fibrous matrix. Solid particles could be, e.g., sublimated or dissolved in a technological liquid and washed out, and drops and liquid films precipitated on fibers can be evaporated or extracted by other liquids. FEs employed in air cleaning in salt mines, drying compressed air, or water removal from benzene, can undergo these procedures.

Regeneration of melt-blown sheet sorbents intended for petroleum products is performed by squeezing and washing with hot water. Upon oil removal, sorbents of SXT, UXT grades, meshes, and seines of the SPC Co. can be used repeatedly. The company also offers extractors – special equipment for sorbent regeneration and isolation of liquids collected by them.

Utilization means use with profit and presupposes repeated use of materials or articles after they are work out. The main types of melt-blown material use are recycling, incineration, pyrolysis, and depolymerization.

Recycling is the most effective method of using worn out products. With the help of a special processing technique, the product is transformed into a recycled material. Recycling of used up polymer products gives rise to a number of problems which involve gathering, sorting, and primary processing of wastes; selection of processing methods and modification of recyclable materials; development of a range of articles produced from waste, and so on. Experience in reprocessing plastic products has specified the structure and methods of returning the wastes of melt-blown materials into the production sphere.

Technological wastes derived in the polymer melt blowing process are rather negligible. They are mainly produced either at start-up and stoppage of equipment or when some disorders in spraying, cutting, or molding of sheets, cartridges, or some other melt-blown products occur. Losses in fiber and modifier spraying can be completely eliminated by controlling the spray jet and forming substrate dimensions, by optimum sealing of the spray chamber, and other methods described in Chap. 2. This is why the total amount of technological wastes does not exceed 5%. Wastes of the melt blowing process differ essentially in molecular mass and physical and mechanical characteristics from the starting polymer due to process specificity (the high temperature of the sprayed melt and intensive mechanical destruction and oxidation of the fiber material). Adding up to 15% of recycled material to the initial raw polymer is recommended.

Production wastes are accumulated with time as melt-blown articles break down. Among all other types of wastes, melt-blown wastes are the most uniform and so present great interest from the viewpoint of recycling them.

A typical recycling technology for worn out melt-blown products usually includes initial sorting and identification of articles by their fiber material, followed by milling, grading of mixed wastes, washing, drying, and granulation, if necessary.

Fibrous articles are cut into fragments 5–30 mm in size using rotary knife mincers. Milling of polymer wastes presents a serious engineering problem of achieving a uniform supply of the fibrous mass in the cutting zone, timely sharpening of the knives with minimum stoppage, reducing the energy requirements of the process, and so on. Original designs of highly efficient mincing machines for milling fibrous materials have been elaborated at MPRI [8]. A revolving knife block of the mincer is aligned with a stationary knife installed inside the drum housing on which milled fibers are wound. Knives of the revolving block are brought into contact with the stationary knife and enter into ports in the wall of the drum rotating synchronously with the knife block [9]. Such mincers contain abrasive tools for knife sharpening [10] and devices to regulate clearances between the stationary and rotating knives

[11, 12] which can be set automatically to adjust the clearances, depending on the regime of material supply to the milling zone [13]. The cutting edges of knives are made in a form of generatrix of a one-pole hyperboloid of revolution [14]. The degree of atomizing of melt-blown product defines the volume density and friability of secondary material, as well as the necessity for granulating it.

During waste milling and further grading (flotation, aero- and electrical separation, chemical methods, etc.), the fibrous base of melt-blown products can be separated from modifiers and entrained by FM solid contaminants.

The expediency of granulating milled melt-blown wastes is determined by the method of further processing, depending on whether they are treated as an independent product or in mixture with the initial polymer material, and on the latter's share in the mixture. The granulation efficiency of plastic wastes by standard extrusion equipment is comparatively high and depends on the type of recyclable thermoplastic and its rheological characteristics (for example, it consists of 40–50 kg/h for melt-blown LDPE).

The kind and prehistory of used articles dictate the purpose and method for recycling them. Recycled materials obtained from unmodified melt-blown products and those containing inert components are reprocessed either in a mixture with initial polymers or separately. Inert modifiers and contaminants captured by FM play the role of fillers in secondary materials. Recycled materials produced from magnetic melt-blown articles are used to impart a weak magnetic field source to a final product. The electret charge is relaxed when recycling melt-blown fibers. Oil sorbing materials and goods are reprocessed into plasticized secondary materials. Used up filtering materials containing concentrated hazardous contaminants in the solid or liquid phase are not subjected to recycling. It is worthwhile to reprocess melt-blown biomass carriers into biodegradable films and packing materials. Problems of recycling these and some other types of melt-blown materials are discussed following.

Domestic wastes are accumulated in dwelling areas, at catering establishments, public facilities, and etc. Such wastes in urban dumps turn into a category of "mixed wastes" which comprise more than half of all waste materials. The share of melt-blown wastes varies in different countries between 1 and 10%.

Processing mixed wastes presents much difficulty necessitating stage by stage sorting of various types into categories. Gathering of used up polymer and, in particular, melt-blown materials, is hard to organize, so in many countries it is not yet launched. That is why the amount of domestic waste recycled into renewed products is in fact nil.

Incineration of plastic wastes is most often performed to use heat generated in burning to produce vapor and electrical power. However, the heat value of such a fuel, especially a melt-blown one, is rather low; therefore, power plants operating on it are, as a rule, inefficient. Besides, toxic gases generated by plastics incineration pollute the atmosphere. In addition, incomplete combustion of polymer wastes is accompanied by soot formation leading to rapid corrosion of furnaces.

Pyrolysis and depolymerization processes deriving monomers or oligomers are ecologically safer alternatives to plastic waste incineration.

Pyrolysis includes temperature-induced (800–1000°C) decomposition of substances. It is considered one of most productive methods of regenerating plastic wastes and is widely applicable all over the world. Melt-blown waste is transformed through pyrolysis into highly efficient fuel, and also into raw materials and semiproducts intended for the chemical industry. Recently, catalytic hydrocracking processes have been elaborated to produce benzene and fuel oils from polymer refuse. PE wastes are decomposed thermally without oxygen access to obtain activated and technical carbon wanted in numerous industrials sectors. Hazardous fiber modifiers released during pyrolysis of melt-blown wastes, as well as heavy and radioactive metal particles, undergo special treatment after which they can be reused in small-tonnage chemistry.

Depolymerization converts melt-blown wastes into low molecular weight saturated hydrocarbons. The latter are then subjected to cracking to a yield of unsaturated compounds used afterward in polyolefin synthesis. So, melt-blown wastes that have a lower molecular mass in contrast to starting thermoplastics decompose giving monomers at comparatively low temperatures and have proved to be an optimum raw material for depolymerization.

Burial of wastes has become a global problem at the border of twenties and twenty-first centuries [15]. The strategy of engineering ecology in this respect can be characterized as follows today. In human history, the main portion of refuse is either buried in the earth or incinerated. At present, a number of countries have adopted laws in the sphere of ecology, which limit inefficient, hard to implement, and hazardous ways of waste disposal. To obtain official permission to build a modern setup for refuse incineration is almost impossible in the majority of countries. Communities are extremely anxious now about air purity and are eager to diminish atmospheric contamination by fumes and waste gas emissions. An overwhelming part of wastes replenishes dumps and constitutes 90% in Europe, and a bit less in North America thanks to advances in recycling programs. Domestic and ecologically safe refuse is transported to great distances outside residential areas and finding new places for dumps is becoming practically impossible day by day. The storage of hostile wastes is highly expensive and will multiply in the near future. The European community is going to adopt even tougher standards for waste burial methods and dump projects.

Burial in the ground is the most widespread method of waste disposal despite the fact that polymer assimilation in soil proceeds for 70–80 years [16], and dumps occupy farmlands.

The specificity of melt-blown materials as biodegradable objects of burial is in their fibrous structure and the presence of an active surface layer on the fibers. The layer, as has been shown previously, contains a great number of free radicals and unbalanced oxygen- containing groups. Along with this, melt-blown materials are a source of weak electrical fields, which stimulate microorganism multiplication and vitality at optimum intensity [17]. This is the reason why melt-blown materials degrade biologically much faster than common polymer waste. In this respect, they resemble bio-, photo- and water-soluble polymers, work on which was gaining strength at the beginning of 1970s. The development of such materials has taken the following routes: (1) introduction into macromolecules of functional groups that can decompose under UV rays and aerobic and anaerobic microorganisms; and (2) filling polymer binders with biological matter that serves as a culture medium for microorganisms [18]. It was soon found that the high physicomechanical characteristics of structural plastics were hard to match with a nice appearance and preset service life without quality impairment. Nevertheless, this idea can be applied well and extended to melt-blown materials because requirements for their external appearance and strength are not so rigid. Progress in this direction could give rise to a new generation of filtering, medical, and disposable materials whose biodegradation does not present any problem.

12.3 Economic Estimates

Beginning with the 1970s, melt-blown materials have occupied a leading place in the world's production of fibrous materials immediately after Exxon Co. started to sell licenses for melt blowing technology. In the 1960s, the use of melt-blown materials was limited just to separators of chemical current sources, whereas 20 years later, practically all classes of fibrous materials have been enriched by melt-blown analogues followed by elaboration of novel types of commercial products on a melt-blown base.

Commercialization of the melt blowing technique was a significant event in the international development of the industrial structure with far-reaching outcomes:

- the number of operations in fibrous material production cycle reduced (spinning was eliminated, trimming and fiber tying were simplified, etc.);
- the productivity of textile equipment increased owing to combined operations of molding, modification and tying, canceling thread cutoff, as well as fiber strengthening by orientation extension;
- the material consumption of fibrous articles reduced through fiber thinning during the melt blowing process;
- the assortment of textiles extended thanks to the diversity of chemical and physical treatments of the sprayed melt fragments, and melt blowing combinations with other processes;
- the service characteristics of fibrous products improved, they have become multifunctional through control of shape stability and fibrous carcass

rigidity and serve as electrical and magnetic field sources, as well as carriers for microorganisms, adsorbents, and other target additives.

Filtration has become one of the main fields for application of melt-blown articles because melt blowing furnishes wide technological means to (1) control the porosity and strength of a FE, (2) mold and original FE of practically any shape having a porosity gradient and that of functional components distribution, (3) modify FM by physical fields and chemical and biological procedures, (4) stimulate the growth and active life of microorganisms settled on fiber surfaces, and (5) realize various mechanical and physicochemical mechanisms of contaminant entrapment in a melt-blown FE. The appearance of technically perfect means of filtration has made it possible in recent decades to solve most advantageously a wide circle of problems in engineering ecology and to increase the life of equipment.

Melt-blown materials have entered a range of construction goods first as exotic decorative elements of buildings and then as irreplaceable heat and sound insulating boards, waterproofing facings, and multifunctional components of floors. Modern medicinal technique is impossible to imagine without vessel endoprostheses with melt-blown elements, novel generations of dressing materials convenient for medical staff and patients, bandages, splints, and new curing remedies for burns which are highly efficient due to melt-blown components.

Unconventional structures and shapes of certain products on a melt-blown base (chemical current sources, electrical insulation, packages, etc.) form a totality of contemporary technical goods that have superlative performance.

The role of melt-blown materials, which compete with or/and supplement traditional fabric, leather, and carpet goods, is to be emphasized separately. Their unusual texture, color, and attractive appearance have enriched textile and light industry potentialities in creating super comfortable, fine, competitive, and efficient products.

The modern world uses a broad range of melt-blown disposables (napkins, diapers, towels, clothes, cleaning and polishing agents) which constitute a separate production branch.

The dynamic advancement in production and consumption of melt-blown materials and products in all countries have led to a substantial renewal of their range and affected the international economy profoundly.

Beginning with the 1970s, the global production of melt-blown articles has grown rapidly at a 80–100% yearly increment in production volume far surpassing that of textile fiber manufacture. Geographically the melt-blown industry is concentrated largely in the U.S., Japan, and European countries. As for CIS states, a greater portion of melt-blown goods is produced by small ventures, and the quality of their production depends on the technical level and industrial engineering at an enterprise.

The cost pattern of the basic productive assets of the melt-blown goods industry approaches that of chemical fibers, allotting 45% to buildings and structures, and 49% to machines and equipment. Within the pattern of expenses for producing melt-blown products, about 50% goes to raw materials, 20% for power charges, and about 10% for salaries.

The gradual depletion of oil, coal, and gas resources along with the growing complexity of their extraction and the increase in energy cost are the reason why the starting polymer materials and, consequently, melt-blown articles are becoming more and more expensive. The way out is upgrading the productivity of the melt blowing technique and increasing production volume. Melt-blown products with a minimum prime cost have a polyolefin base; the maximum costs are for articles based on polymers that form highly viscous melts at a temperature below their thermally induced oxidative destruction temperature, i.e., high molecular mass polymers with a large amount of active side groups (high molecular weight PE, polyesters, polyarylates, and so on).

Production of melt-blown products is one of most profitable plastics processing areas today. The ratio of income to productive assets is 25–27% in this sphere. The benefits of further progress here are conditioned by a satisfactory state of the raw materials base, advantages in contrast to more costly alternative products based on synthetic fibers, and vast markets for diverse melt-blown materials and products.

13. Conclusion

Little known so far to a customer at large melt-blowing technique is now seriously and for long holding its place among a totality of production technologies being the base of modern industry. Recently elaborated numerous melt-blown materials and articles have speeded up the development of a number of major fields of human activities (industry, construction, engineering ecology, etc.) and are rightfully associated with the contemporary level of civilization.

This science-consuming technique will, obviously, be intensively advancing by utilizing fundamental knowledge in physics, chemistry and mechanics of contact interactions between solid, liquid and gas and tending to interlace with biology and medicine. Joint efforts of experts in various domains are anticipated to impact the search of the following new directions in melt-blowing process.

A novel class of polymer materials is expected to be created specifically for melt-blowing procedure which will display high flow at processing temperatures avoiding macromolecular destruction. Groups of easily sprayed blend composites, new combinations of plastic components serving simultaneously to regulate molten binder viscosity are to be developed soon.

Refined melt-blowing equipment will allow composite plastic processing and forming of combined parts that incorporate metals, ceramics and plastics. Break-through in designing extrusion heads is also anticipated as a result of substitution of costly multihole heads by more simple and highly efficient slot heads and improved spraying quality.

New application fields of melt-blowing technique will be open by the chance to regulate cohesive interaction and spreading of deposited on substrate fibers. Wide range of structural and technological characteristics improvement is furnished by the transfer from traditional fibrous materials to pore-free plastics produced from pressed melt-blown half-finished products. This opens new possibilities to upgrade service characteristics of melt-blown articles intended for triboengineering, sealing and anticorrosion technique.

Up till now melt-blowing was mainly directed to processing thermoplastics into fibrous materials with various degree of fiber bonding. It could be anticipated that melt-blown materials will tend to be basically multifunctional composites intended not only for engineering, but for biology, medicine and other spheres of human activity too.

Engineering ecology has become a branch of using melt-blown materials on a large scale. The structure of these materials facilitates interactions of natural phenomena, such as moisture permeation, diffusion of organic compounds, microorganism growth, and so on. The nearest prospect in this field is expected to be price cut of melt-blown means for engineering ecology by processing recyclable polymers and development of composites on their base with regulated biocompatibility and biodegradability degree.

Biotechnology has contributed much into solution of global problems being a spacious proving ground for melt-blown materials testing as microorganism carriers. Named materials are water and air permeable, their electric field effects surrounding and stimulates multiplication and active life of microorganisms. Todays melt-blown materials outdo all other known engineering materials in properties specifying them as microorganism habitat.

A powerful impetus on improving quality of melt-blown materials and refining equipment designs can make the development of a physics-chemical model and methods of describing melt-blowing process using representations of the theory of oxidation, aero-, hydro and thermodynamics. The familiar models of liquid spraying will be thus modified to consider specificity of overheated polymer melt flowing in a viscous-flow state through extrusion head spinnerets and peculiarities of a compressed gas or liquid spraying kinetics.

So, prospects in advancement of melt-blown materials are connected with imparting additional functional properties to them. Along with their major designation in one or another sphere, they will also fulfill auxiliary functions of being carriers of electric or magnetic fields, bactericides or corrosion inhibitors, and so on. The creation of such kind materials is, as a rule, connected with overcoming preconceptions concerning probability of their use while their technological processing, formulations and areas of optimum application are characterized by the novelty and competitiveness. Although traditional melt-blown materials are still keeping their importance, search for new raw material sources and their rational processing methods into composites on melt-blown base is being under way. The latter are enriching the generation of smart materials whose structure and properties are prone to regulation under varying operation regimes.

As can be seen, together with traditional representations on the family of melt-blown materials the authors tried to reflect unacknowledged universally but being in accord with tending to refinement melt-blown production concepts. The authors hope that the book will be a stimulus for fresh investigations in the field of polymer fibrous materials.

References

Chapter 1

1. A. Ziabicki. *Fundamentals of Fiber Formation*. London, Wiley, 1976.
2. S.P. Papkov. *Polymer Fibrous Materials*. Moscow, Khimia, 1986.
3. J. Patockova. Tvorba rouna rozfukovanim taveniny termoplastickeho polymeru. *Textil*, 1989, Vol. 44, No. 7, pp. 250–253.
4. M.T. Gillies. *Nonwoven Materials.* New Jersey, Noyes Data, 1979.
5. U.S.S.R. Patent 278478, B 44 D 1/08. A method for plastic coating application. V.A. Belyi, publ. 1970.
6. V.P. Shustov. Elaboration and investigation of new methods for producing engineering materials and coatings on polymer base. Ph.D. Thesis, Minsk, 1975.
7. T.V. Stavrova. Elaboration and investigation of fibrous-porous materials produced by polymer melt spraying. Ph.D. Thesis, Minsk, 1983.
8. V.P. Shustov, T.V. Stavrova, and O.R. Yurkevich. On the law of particle distribution by size at polymer melt spraying. *News BSSR AS, Ser. Phys.-Tech. Sci.*, 1976, No. 3, pp. 126–129.
9. R. Beyreutler, H. Brunig, and R. Vogel. Melt spinning (polymer formation under high stress conditions). *Polymer Materials Encyclopedia*, Vol. G. Boca Raton, Florida, CRC Press, 1996, pp. 4061–4074.
10. A.V. Genis. Investigation of fibrous material formation from polyamide melt. *Chem. Fibers*, 1998, No. 2, pp. 25–28.
11. I.N. Ermolenko, T.M. Uliyanova, P.A. Vityaz, and I.L. Fedorova. *Fibrous High-Temperature Ceramic Materials*. Minsk, Nauka i Tekhnika, 1991.

Chapter 2

1. U.S. Patent 4526733, B 29 F 3/04. Melt-blown die and method. J.C. Lau, 1985.
2. U.S. Patent 4032688, B 29 C 47/30. Process for forming nonwoven thermoplastic fibrous material. D.B. Pall, 1979.
3. M.T. Gillies. *Nonwoven Materials*. New Jersey, Noyes Data, 1979.
4. U.S.S.R. Patent 387570, B 44 D 1/08. A method to spray polymer melts. V.P. Shustov, V.A. Belyi, and V.P. Stavrov, 1973.
5. U.S. Patent 4797318, D 04 H 1/58. Active particle-containing nonwoven material, method of formation thereof, and uses thereof. R.W. Brooker, B. Cohen, and D.M. Jackson, 1989.
6. U.S.S.R. Patent 690685, B 05 B 7/06, B 05 D 1/02. A method of pneumatic spraying a thermoplastic melt. V.A. Belyi, V.P. Shustov, and T.V. Stavrova, 1980.

7. U.S. Patent 4380570, D 04 H 1/4. Apparatus and process for melt blowing a fiberforming thermoplastic polymer and product produced thereby. E.C.A. Schwarz, 1983.
8. U.S. Patent 4847125, F 16 L 11/02. Tube of oriented, heat shrunk, melt blown fibers. E.C.A. Schwarz, 1989.
9. U.S.S.R. Patent 1632102, D 04 H 3/00. A method to produce porous material from a polymer melt. F.D. Sidorenko, V.F. Gaiduk, A.V. Sikanevich, and V.P. Shustov, 1991.
10. U.S.S.R. Patent 427573, B 44 D 1/08, C 09 D 3/48. A method of coating formation. V.A. Belyi, V.P. Shustov, and O.R. Yurkevich, 1974.
11. Belarus Patent 1484, B 29 C 41/08. A procedure to produce filtering materials. L.S. Pinchuk, A.V. Makarevich, V.A. Goldade, et al., 1996.
12. Belarus Patent 1481, B 29 C 41/08. A procedure to produce filtering materials. L.S. Pinchuk, A.V. Makarevich, E.I. Parkalova, et al., 1996.
13. Belarus Patent 1810, B 29 C 41/08. A method to obtain fibrous nonwoven materials. L.S. Pinchuk, A.V. Makarevich, Yu.V. Gromyko, et al., 1997.
14. Belarus Patent 1424, B 29 C 41/08, D 04 H 1/56. A method to produce fibrous structural elements and a device for its implementation. L.S. Pinchuk, M.S. Semenyuk, V.A. Goldade, et al., 1996.
15. U.S.S.R. Patent 1434737, C 08 J 5/18. A method to obtain filtering material from polymer fibers. Yu.V. Gromyko, V.S. Mironov, L.S. Pinchuk, et al., 1988.
16. Belarus Patent 2340, B 01 D 35/06, 39/16. A filtering material and method of its production. L.S. Pinchuk, Yu.V. Gromyko, V.A. Goldade, et al., 1998.
17. Belarus Patent 2803, B 01 D 39/16. A method for filtering material production. L.S. Pinchuk, A.V. Makarevich, and A.G. Kravtsov, 1999.
18. Belarus Patent 2571, B 22 F 3/20, H 01 F 1/113. A method to manufacture articles from a magnetic polymer material. L.S. Pinchuk, A.V. Makarevich, and A.G. Kravtsov, 1998.
19. U.S.S.R. Patent 586628, B 29 D 27/00. A method to form porous articles from polymer melt. V.P. Shustov, V.A. Belyi, and T.V. Stavrova, filed December 22, 1975.
20. U.S.S.R. Patent 612470, B 29 D 7/00. A method of manufacturing sheet polymer material. V.P. Shustov, V.A. V.A. Belyi, T.V. Stavrova, and O.R. Yurkevich, filed May 16, 1975.
21. U.S.S.R. Patent 1066120, B 29 D 23/00. A method to manufacture drain-pipe. N.G. Pivovar, V.P. Shustov, T.V. Stavrova, et al., filed May 20, 1982.
22. U.S.S.R. Patent 1578948, B 29 C 41/08. A method to produce porous articles from polymer material. F.D. Sidorenko, A.V. Sikanevich, V.P. Shustov, and V.F. Gaiduk, filed July 11, 1988.
23. U.S.S.R. Patent 1773881, C 02 F 3 /20. An aerating element. F.D. Sidorenko, A.V. Sikanevich, V.F. Gaiduk, and V.P. Shustov, 1992.

Chapter 3

1. U.S. Patent 4818464, B 65 H 54/00. Extrusion process using a central air jet. J.C. Lau, 1989.
2. U.S.S.R. Patent 539615, B 05 B 1/26. A spraying device. V.A. Belyi, A.N. Belyaev, O.M. Shalobalov, et al., 1976.
3. U.S.S.R. Patent 1419739, B 05 B 7/26. A pneumatic sprayer. V.P. Shustov, T.V. Stavrova, F.D. Sidorenko, and V.A. Gorbachenko, 1988.

4. U.S.S.R. Patent 388918, B 44 D 1/08. A device for coating application. V.P. Shustov, V.A. Belyi, T.V. Stavrova, and V.P. Stavrov, 1973.

5. U.S.S.R. Patent 1209306, B 05 B 3/12. A device for polymer melt spraying. V.P. Shustov, T.V. Stavrova, and V.A. Gorbachenko, 1986.

6. U.S.S.R. Patent 699055, D 04 H 3/00. A device for polymer melt spraying. V.P. Shustov, T.V. Stavrova, and N.M. Klimashevich, 1979.

7. Poland Patent 161651, C 03 B 37/06. A device for plastics dispersion. A. Patrin, J. Dragula, B. Nadgorsky, et al., 1993.

8. U.S. Patent 3825380, B 29 F 3/04. Melt blowing die for producing nonwoven mats. J.W. Harding, J.P. Keller, and R.R. Buntin, 1972.

9. U.S. Patent 4720250, B 28 B 5/00. Slotted melt-blown die head. D.W. Appel, A.D. Drost, and J.C. Lan, 1988.

10. U.S. Patent 4986743, B 29 C 47/30, D 01 D 5/08. Melt blowing die. P.G. Buehning, 1991.

11. U.S. Patent 4526733, B 29 F 3/04. Melt-blown die and method. J.C. Lau, 1985.

12. U.S. Patent 5196207, B 29 C 47/12. Melt-blown die head. R.J. Koenig, 1993.

13. U.S. Patent 4380570, D 04 H 1/04. Apparatus and process for melt blowing a fiber-forming thermoplastic polymer and product produced thereby. E.C.A. Schwarz, 1983.

14. U.S. Patent 5476616, B 29 C 47/30. Apparatus and process for uniformly melt blowing a fiber-forming thermoplastic polymer in a spinneret assembly of multiple rows of spinning orifices. E.C.A. Schwarz, 1995.

15. Belarus Patent 1602, B 29 C 47/30. Extrusion head. L.S. Pinchuk, M.S. Semenyuk, V.A. Goldade, et al., 1997.

16. Belarus Patent 2021, B 29 C 47/30. Extrusion head. L.S. Pinchuk, V.A. Goldade, and A.V. Makarevich, 1998.

17. Belarus Patent 2341, B 29 C 47/30. Extrusion head. L.S. Pinchuk, A.V. Makarevich, and E.I. Parkalova,1998.

18. Belarus Patent 3514, B 29 C 47/30. Filtering material. L.S. Pinchuk, A.V. Makarevich, and A.G. Kravtsov, 2000.

19. Belarus Patent 3229, B 29 C 47/30. Extrusion head. L.S. Pinchuk, A.V. Makarevich, V.A. Goldade, and V.G. Plevachuk, 2000.

20. V.A. Belyi, V.A. Goldade, A.S. Neverov, and L.S. Pinchuk. Electret state of polymers in splices of inhomogeneous metals. *Proc. USSR AS*, 1979, Vol. 245, No. 1, pp. 132–134.

21. Belarus Patent 1617, B 29 C 47/20, 47/26. An extrusion head for hose polymer material production. L.S. Pinchuk, A.V. Makarevich, V.A. Goldade, and Yu.V. Gromyko, 1997.

22. Belarus Patent 2996, B 29 C 47/30. Extrusion head. L.S. Pinchuk, A.V. Makarevich, V.A. Goldade, and A.G. Kravtsov, 2000.

23. Melt-Blown Production. Advert. Booklet of Biax–Fiberfilm Corp., U.S., 1997.

24. U.S.S.R. Patent 921183, B 29 C 5/04. A device for manufacturing porous articles from polymer melts. V.P. Shustov, O.M. Shalobalov, V.M. Sukhorukov, and T.V. Stavrova, 1982.

25. Belarus Patent 1424, B 29 C 41/08, D 04 H 1/56. A method to produce structural elements and a device for implementation. L.S. Pinchuk, M.S. Semenyuk, V.A. Goldade, et al., 1996.

26. U.S. Patent 4968238, B 29 C 47/92, 71/04. Apparatus for making a non-woven sheet. R.A. Satterfeld and D.M. Taylor, 1990.

27. Belarus Patent 418, B 29 C 41/08 and 67/20, B 29 K 105/04, B 29 L 23/00. A device to mold pipes from thermoplastic material. F.D. Sidorenko, A.V. Sikanevich, V.F. Gaiduk, and Yu.I. Gorbachev, 1994.

28. U.S. Patent 4847125, F 16 L 11/02. Tube of oriented, heat shrunk, melt-blown fibers. E.C.A.Schwarz, 1989.

29. U.S. Patent 4223059, B 29 C 17/02, D 04 H 3/12, D 06 C 3/06. Process and product thereof for stretching a non-woven web of an orientable polymeric fiber. E.C.A. Schwarz, 1980.

30. Belarus Patent 1592, B 29 C 41/08, D 04 H 1/56. A device to produce fibrous nonwoven materials. L.S. Pinchuk, A.V. Makarevich, and V.A. Goldade, 1997.

31. R. Beyreutler, H. Brunig, and R. Vogel. Melt-spinning (polymer formation under high stress conditions). *Polymeric Materials Encyclopedia*, Vol. 6, Boca Raton, Florida, CRC Press, 1996, pp. 4061–4074.

32. Germany Patent Application 3744657, D 04 H, D 01 D 5/088. A method and apparatus for spinning nonwoven material from synthetic fibers. V. Andreevsky, M. Honke, and K. Max, 1988.

33. U.S. Patent 5186879, B 29 C 47 / 88. Spinning process for producing high strength, high modulus, low shrinkage yarns. F.H. Simons and R.L. Griffith, 1993.

Chapter 4

1. V.G. Plevachuk, A.V. Makarevich, E.I. Parkalova et al. Structural and adsorption characteristics of nonwoven fibrous polymeric filtering materials. *Chem. Fibers*, 1997, No. 1, pp. 31–34.

2. V.A. Goldade, L.S. Pinchuk, A.V. Makarevich, and V.G. Plevachuk. Research of structure and adsorption characteristics of polymeric fibrous filtering materials. *The 12th Annu. Meeting Polym. Process. Soc.*, Sorrento, 1996, Extended Abstracts, pp. 307–308

3. S.A. Saltykov. *Stereometrical Metallography*. Moscow, Metallurgija, 1970.

4. M. Nadler, and E.P. Smith. *Pattern Recognition Engineering*. NY, John Wiley & Sons, 1993.

5. T.B. Kirk, D. Panzera, R.V. Anamalay, and Z.L. Xu. Computer image analysis of wear debris for machine condition monitoring and fault diagnosis. *Wear*, 1995, Vol.181–183, pp. 717–723.

6. S.K. Akhnazarova, and V.V. Kafarov. *Experiment Optimization in the Chemical Industry*. Moscow, Vyshaja shkola, 1978.

7. L.S. Gradus *Instruction on Dispersion Analysis by the Microscopic Method*. Moscow, Khimija, 1979.

8. M.M. Dubinin. Surface and porosity of adsorbents. *Chem. Rev.*, 1982, Vol. 51, No. 7, pp. 1065–1074.

9. A.W. Adamson. *Physical Chemistry of Surfaces*, 3rd ed., Los Angeles, California, John Wiley & Sons, 1976.

Chapter 5

1. L.I. Tarutina, and F.O. Pozdnyakova. *Spectral Analysis of Polymers*. Moscow, Khimia, 1986.

2. L.S. Pinchuk, A.G. Kravtsov, Yu.I. Voronezhtsev, and Yu.V. Gromyko. On the charge state of melt-blown polymer materials, *Int. J. Polym. Process.*, 1998, Vol. 13, No.1, pp. 67–70.

3. A.I. Kuzavkov. Adhesion strength of polyethylene coatings oxidized in diffusion regime on steel and aluminum. Ph.D. Thesis, Kiev, 1983.

4. N.I. Egorenkov. Regularities in changing of adhesive, frictional and physico-chemical properties of polymer films on metals under thermal actions. Dr. Sci. Thesis, Kiev, 1987.

5. D.G. Lin. Contact oxidation influence of polyethylene coatings on adhesion to metals. Ph.D. Thesis, Moscow, 1973.

6. V.A. Belyi, N.I. Egorenkov, and Yu.M. Pleskachevskij. *Adhesion of Polymers to Metals*. Minsk, Nauka i Tehnika, 1971.

7. N.I. Egorenkov, D.G. Lin. and V.A. Belyi. Contact oxidation of polyethylene melt on metal alloys. *Proc. of USSR AS*, 1974, Vol. 214, No. 6, pp. 1376–1379.

8. N.I. Egorenkov, D.G. Lin, and V.A. Belyi. Influence of polyethylene melt contact reactions with metal on polyethylene oxidation, *Proc. of USSR AS*, 1972, Vol. 207, No.2, pp. 397–400.

9. V.I. Nefedov. *X-ray Electron Spectroscopy of Chemical Compounds*. Moscow, Khimia, 1984.

10. D. Briggs and M.P. Seach, eds. *Practical Surface Analysis by Auger and X-ray Photoelectron Spectroscopy.* Chichester, NY, Brisbane, Toronto, Singapore, John Wiley & Sons, 1983.

11. L.A. Kazitsyna, and N.B. Kupletskaya. *UV-, IR- and NMR-Spectroscopy in Organic Chemistry*. Moscow, Vyschaya shkola, 1971.

12. A.J. Gordon, and R.A. Ford. *The Chemist's Companion. A Handbook of Practical Data, Techniques, and References*. NY, London, Sydney, Toronto, John Wiley & Sons, 1972.

13. K. Nakamoto. *Infrared and Raman Spectra in Organic and Coordination Compounds*. Brisbane, Toronto, Singapore, John Wiley & Sons, 1986.

14. K.E. Lawson. *Infrared Absorption of Inorganic Substances*. NY, Reinhold, 1961.

15. K.S. Kao, and W. Hwang. *Electrical Transport in Solids*. Oxford, NY, Toronto, Sydney, Paris, Frankfurt, Pergamon Press, 1981.

16. G.M. Sessler, ed. *Electrets*. Berlin, Heidelberg, London, NY, Tokyo, Springer-Verlag, 1987.

17. A.G. Milnes. *Deep Impurities in Semiconductors*. NY, Wiley–Interscience, 1973.

18. J. van Turnhout. *Thermally Stimulated Discharge of Polymer Electrets*. Amsterdam, Elsevier, 1975.

19. A.G. Kravtsov. Development of polymeric fibrous magnetic materials for fine purification of technological media. Ph.D. Thesis, Gomel, 1998.

20. G.M. Bartenev, and R.M. Aliguliev. Relaxation spectrometry of low density polyethylene, *High–Mol. Compounds*, 1982, Vol. 24 A, No. 9, pp. 1842–1849.

21. Belarus Patent 2340, B01D 35/06, 39/16. Filtering material and method of its manufacturing. L.S. Pinchuk, Yu.V. Gromyko, V.A. Goldade, et al. 1998.

Chapter 6

1. State Standard 14146-88 Russia. Filters for cleaning diesel fuel. General Technical Specifications.

2. State Standard 25277-82 Russia. Filtering elements for large-size hydraulic drives and lubrication systems. Rules of acceptance and testing methods.

3. State Standard 25476-82 Russia. Large-size hydraulic drives and lubrication systems. Filters. Rules of acceptance and testing methods.

4. State Standard 26070-83 Russia. Filters and separators for liquids. Terms and definitions.

5. C. Dickenson, ed. *Filters and Filtration*, 3rd ed. Oxford, Elsevier, 1992.

6. O.S. Balabekov, and L.Sh. Baltabaev. *Gas Cleaning in the Chemical Industry*, Moscow, Khimia, 1991.
7. U.S. Patent 4847125, F16L 11/02. Tube of oriented, heat shrunk, melt-blown fibers. E.C.A. Schwarz, 1989.
8. German Patent 3413213, B01D 39/14. Filtering element, H. Walter, 1985.
9. J. van Turnhaut. The use of polymers for electrets. *J. Electrostatics*, 1975, No. 1, pp. 147–163.
10. P.A. Kousov, A.D Malygin, and G.M. Skryabin. *Gas and Air Purification in the Chemical Industry*, Leningrad, Khimia, 1982.
11. V. Straus. *Industrial Gas Cleaning*, Moscow, Khimia, 1981.
12. H.J. White. *Industrial Electrostatic Precipitation.* Reading, Massachusetts, Addison–Wesley, 1963.
13. E.R. Frederick. How dust filter selection depends on electrostatics. *Chem. Eng.*, 1961, Vol. 68, No. 13, pp. 107–114.
14. A.D. Zimon. *Adhesion of Dust and Powders*, Moscow, Khimia, 1967.
15. W.R. Harper. *Contact and Frictional Electrification*, Oxford, 1967.
16. A.V. Sanduliak. *Magnetic Filtration Cleaning of Liquids and Gases*, Moscow, Khimia, 1988.

Chapter 7

1. R.C. Brown. *Air Filtration: An Integrated Approach to the Theory and Applications of Fibrous Filters.* Oxford, NY, Seoul, Tokyo, Pergamon Press, 1993.
2. C.N. Davies. *Air Filtration.* London, NY, Academic Press, 1973.
3. G.M. Sessler, ed. *Electrets.* 2nd ed. Berlin, Heidelberg, NY, London, Paris, Tokyo, Springer–Verlag, 1987.
4. J. van Turnhout. The use of polymers for electrets. *J. Electrostatics*, 1975, No.1, pp. 147–163.
5. Yu.V. Gromyko, and A.G. Kravtsov. Electret state of fibrous material based on polyethylene. *Proc. Belarus AS*, 1995, Vol.39, No.5, pp. 112–116.
6. G.M. Bartenev, and R.M. Aliguliev. Relaxation spectrometry of low density polyethylene. *High-Mol. Comp.* 1982, Vol. 24A, No.9, pp. 1842–1849.
7. J. van Turnhout. *Thermally Stimulated Discharge of Polymer Electrets.* Amsterdam, Elsevier, 1975.
8. K.C. Kao, and W. Hwang. *Electrical Transport in Solids.* Oxford, NY, Toronto, Sydney, Paris, Frankfurt, Pergamon Press, 1981.
9. J. Frenkel. On pre-breakdown phenomena in insulators and semiconductors. *Phys. Rev.*, 1938, Vol. 54, pp. 647–648.
10. A.D. Zimon. *Adhesion of Dust and Powders.* Moscow, Khimia, 1967.
11. B.V. Derjaguin, and M.M. Kusakov. An experimental investigation of polymolecular solvate films. *Acta Phys.–Chim. URSS*, 1939, Vol.10, No.1, pp. 25–44; No.2, pp. 153–174.
12. B.V. Derjaguin, and N.V. Churaev. On the question of determining the concept of disjoining pressure. *J. Colloid Interface Sci.*, 1978, Vol.66, No.3, pp. 389–398.
13. B.V. Derjaguin, and L.D. Landau. Theory of the stability of strongly charged lyophobic sols. *Acta Phys.–Chim. URSS*, 1941, Vol.14, No.6, pp. 633–662.
14. A.S. Akhmatov. *Molecular Physics of Boundary Friction.* Moscow, Phismatgiz, 1963.
15. B.V. Derjaguin, and A.S. Titievskaya. Investigations of the forces of interaction of surfaces in different media. *Discuss. Faraday Sos.*, 1954, No.18, pp. 85–98.

16. S. Marchelja, and N. Radich. Repulsion of interfaces due to boundary water. *Chem. Phys. Lett.*, 1976, Vol.42, No.1, pp. 129–134.

17. A.N. Frumkin, V.S. Bagotski, Z.A. Iofa, and B.N. Kabanov. *Kinetics of Electrode Processes*. Moscow, State University, 1952.

18. V.S. Mironov. Electrophysical activation of polymer materials at frictional and electrical affects. Dr. Sci. Thesis, Gomel, 1998.

19. M. Blitshteyn. Wetting tension measurements on corona-treated polymer films. *TAPPI J.*, 1995, Vol.78, No.3, pp. 138–143.

20. W. Lin. and Y.-L. Hsieh. Argon glow discharge and vapor-phase grafting of vinyl monomers on wettability of polyethylene. *Polymer Sci. B: Polym. Phys.*, 1997, Vol.35, No.7, pp. 1145–1159.

21. V.G. Plevachuk, I.M. Vertyachikh, V.A. Goldade, and L.S. Pinchuk. Effect of the charge of polymer electret on the spreading of liquids. *Polymer Sci., Ser. A*, 1995, Vol.37, No.10, pp. 1071–1074.

22. B.D. Summ, and Yu.V. Gorjunov. *Physicochemical Fundamentals of Wetting and Spreading*. Moscow, Khimia, 1976.

23. I.M. Vertyachikh, V.A. Goldade, A.S. Neverov, and L.S. Pinchuk. Influence of electric field of polymer electret on vapor sorption of organic dissolvent. *High–Mol. Comp.*, 1982, Vol.24B, No.9, pp. 683–687

24. I.V. Petryanov, V.I. Kozlov, P.I. Basmanov, and B.I. Ogorodnikov. *Fibrous Filtering Materials FP*. Moscow, Znanie, 1968.

25. Yu.N. Filatov. *Electro-Forming of Fibrous Materials*. Moscow, Oil & Gas Publ., 1997.

26. U.S.S.R. Patent 1434737, C08J 5/18. Method of filtering material manufactured from polymer fibers. Yu.V. Gromyko, V.S. Mironov, L.S. Pinchuk, et al, 1988.

27. A.G. Kravtsov, L.S. Pinchuk, and V.A. Goldade. Melt-blown materials for protecting of respiratory organs. *Chem. Fibers*, 2000, No.6, pp. 42–45.

28. A.I. Sviridenok, A.F. Klimovich, and V.N. Kestelman. *Electrophysical Phenomena in the Tribology of Polymers*. Gordon and Breach, 1999.

29. U.S.S.R. Patent 1351632, B01D 39/00. Filter for aerosols cleaning. Yu.V. Gromyko, A.F. Klimovich, and S.I. Guzenkov, 1987.

Chapter 8

1. L.S. Pinchuk, V.A. Goldade, and Yu.V. Gromyko. Magnetic fibrous polymer materials for ultrafine filtration of liquids. Proc. *6th World Filtration Congr.*, Nagoya, 1993, pp. 940–942.

2. L.S. Pinchuk, and V.A. Goldade. Magnetic polymeric fibrous materials. *Proc. Int. Conf. Adv. Mater. Process. Technol.* (AMPT-93), Dublin, 1993, Vol.1, pp. 347–353.

3. V.A. Goldade, Yu.V. Gromyko, E.M. Markov, and L.S. Pinchuk. Polymeric fibrous magnetic materials for automobile oil filters. *Proc. 26th Int. Symp. Automot. Technol. Automotion* (ISATA-93). Dedicated Conf. on New and Alternative Materials, Aachen, 1993, pp. 391–401.

4. L.S. Pinchuk, V.A. Goldade, and O.K. Kwon. Liquid filtration across fibrous polymer materials – carriers of magnetic field. *Proc. Russian AS*, 1993, Vol.332, No.2, pp. 207–208.

5. V.A. Goldade, and E.M. Markov. Investigation of structure and properties of magnetic fibrous polymer materials. *Mech. Composite Mater.*, 1995, Vol.31, No.3, pp. 291–297.

6. E.M. Markov, L.S. Pinchuk, V.A Goldade., et al. Filtration of wear debris by polymer magnetic filters. *Friction and Wear*, 1995, Vol.16, No.3, pp. 518–522.

7. L.S. Pinchuk, L.V. Markova, Yu.V. Gromyko, et al. Polymeric magnetic fibrous filters. *J. Mater. Process. Technol.*, 1995, Vol.55, pp. 345–350.

8. Kravtsov A.G. Development of polymer fibrous materials for fine filtration of technological media. PhD Thesis, Gomel, 1998.

9. A.V. Sandulyak. *Magneto-Filtration Cleaning of Liquids and Gases.* Moscow, Khimia, 1988.

10. L.V. Markova, E.M. Markov, Yu.V. Gromyko, and L.S. Pinchuk. About liquids filtration across fibrous materials – sources of magnetic fields. *Proc. Belarus AS*, 1994, Vol.38, No.1, pp. 119–122.

11. C.N. Davies. *Air Filtration.* London, NY, Academic Press, 1973.

12. B.G. Ahn, U.S. Choi, O.K. Kwon, and T.J. Moon. Filtration characteristics of fibrous polymeric filters contained magnetic particulate filler. *Adv. Filtration Sep. Technol.*, AFS, 1998, Vol.12, pp. 1–9.

13. L.V. Markova Problems of magneto-optic wear diagnostics of lubricated moving junctions.*Sov. J. Friction Wear*, 1990, Vol.11, No.2, pp. 124–127.

14. C. Dickenson, ed.*Filters and Filtration*, 3rd ed. Oxford, Elsevier, 1992.

15. A.V. Makarevich, A.G. Kravtsov, and L.S. Pinchuk. Influence of spatially-inhomogeneous magnetic fields on coagulation processes in dispersed systems. *J. Appl. Chem.*, 1998, Vol.71, No.5, pp. 817–823.

16. G.A. Luscheikin *Methods of Polymers Electrical Properties Investigation.* Moscow, Khimia, 1988.

17. S.V. Vonsovski. *Magnetism: Magnetic Properties of Dia-, Para-, Ferro-, Antiferro- and Ferri-Magnetics.* Moscow, Nauka, 1971.

18. S. Chikazumi. *Physics of Magnetism.* London, NY, Tokyo, John Wiley & Sons, 1964.

19. O.V. Akopova, and B.V. Eremenko. Stability of quartz water suspensions in electrolyte solutions. *Colloidal J.*, 1992, Vol.54, No.5, pp. 19–23.

20. O.M. Merkushev, A.I. Alekseev, I.S. Lavrov, and A.E. Skachkov. About drops behavior of real emulsions in external electric field. *Colloidal J.*, 1974, Vol.36, No.2, pp. 391–392.

21. N.Ph. Bondarenko, and E.Z. Gak. *Electromagnetic Hydrophysics and Nature Phenomena.* St.-Petersburg, State Agricultural University, 1994.

22. V.I. Klassen. Wettability change of solid bodies by water after magnetic field action. *Proc. USSR AS*, 1966, Vol.166, No.6, pp. 1383–1385.

23. A.M. Demetski, and A.G. Alekseev. *Artificial Magnetic Fields in Medicine.* Minsk, Belarus, 1981.

24. V.I. Klassen. *Magnetization of Water Systems.* Moscow, Khimia, 1982.

25. L.A. Kul'ski, and S.S. Dushkina. *Magnetic Field and Water Treatment Processes.* Kiev, Naukova dumka, 1988.

26. L.S. Pinchuk, E.M. Markov, and A.G. Kravtsov. Magnetic field influence on water flow through clearance of solid bodies contact. *J. Tech. Phys.*, 1996, Vol.66, No.4, pp. 30–35.

27. J.T. Davies, and E.K. Rideal. *Interfacial Phenomena.* NY, Academic Press, 1963.

28. B.D. Summ, and Yu.V. Goryunov. *Physico–Chemical Fundamentals of Wetting and Spreading.* Moscow, Khimia, 1976.

29. Yu.M. Sokol'ski. *Magnetization of Water: The Truth and Fantasy.* Leningrad, Khimia, 1990.

30. F. Franks, ed. *Water: A Comprehensive Treatise.* NY, London, Elsevier,1979, Vol.8.

31. V.I. Minenko. *Electromagnetic Treatment of Water in Heat Power Engineering.* Khar'kov, Prapor,1981.

32. V.I. Yashkevich. About possible mechanisms of outer conditions influence on water systems activation by electromagnetic and other influences. *Abstr. 4th USSR Conf. Magn. Treatment of Water Syst.,* Moscow, State University, 1981, p. 7.

33. O.I. Martynova, B.G. Gusev, and E.A. Leontiev. About mechanism of magnetic field influence on salt water solutions. *Phys. Sci. Rev.,* 1969, Vol.98, No.1, pp. 195–199.

34. E.F. Tabenikchin.*Non-Reagent Methods of Water Treatment in Power Units.* Moscow, Khimia, 1985.

35. Yu.V. Myagkov, and I.V. Myagkov. Model of magnetic activation mechanism. *Abstr. 4th USSR Conf. Magn. Treatment Water Syst.,* Moscow, State University, 1981, pp. 11–12.

36. V. Patrovsky. Hydrogen peroxide in magnetically treated water. *Mol. Phys.,* 1976, Vol. 31, No.4, pp. 1051–1053.

Chapter 9

1. Belarus Patent 2341, B29C 47/30. Extrusion head. L.S. Pinchuk, A.V. Makarevich, and E.I. Parkalova, 1998.

2. Belarus Patent 1810, B29C 41/08. A method to obtain fibrous nonwoven materials. L.S. Pinchuk, A.V. Makarevich, Yu.V. Gromyko, et al., 1997.

3. V.G. Plevachuk, A.V. Makarevich, E.I. Parkalova, et al. Structural and adsorptive characteristics of nonwoven fibrous polymer filtering materials produced by melt-blowing. *Chem. Fibers,* 1997, No 1, pp. 31–34.

4. J. Patockova. Tvorba rouna rozfukovanim taveniny termoplastickeho polymeru. *Textil,* 1989, Vol. 44, No 7, pp. 250–253.

5. Russia Patent 2126715, B01J 20/22. Sorbing fibrous porous material. A.I. Chernorubashkin, A.V. Sikanevich, V.F. Gaiduk, et al., 1999.

6. A.A. Shatov, V.A. Lyubimenko, and V.M. Belyakov. Mathematical model of emulsion filtration in fibrous materials. *Colloidal J.,* 1992, Vol. 54, No 5, pp. 175–181.

7. Belarus Patent 1481, B 29 C 41/08. A procedure to produce filtering materials. L.S. Pinchuk, A.V. Makarevich, E.I. Parkalova, et al., 1996.

8. Belarus Patent 1484, B29C 41/08. A procedure to produce filtering materials. L.S. Pinchuk, A.V. Makarevich, V.A. Goldade, et al., 1996.

9. Belarus Patent 2406, B01D 39/16. Filtering material. A.V. Makarevich, L.S. Pinchuk, V.A. Ostrovsky, et al., 1998.

10. A.V. Makarevich, I.Yu. Ukhartseva, and V.G. Plevachuk. Adsorptive biocidic polymeric filtering materials for water purification. *Polym. Process.Soc.,* Extended Abstr. of Europe/Africa Region Meeting, Gothenburg, Sweden, 1997, Sec.1, Poster 8.

11. L.G. Lavreneva, S.V. Larionov, V.N. Ikorsky, and Z.A. Grankina. Coordinational compositions of transition metals chlorides with tetrazoles. *J. Inorg. Chemistry,* 1985, Vol. 30, No 4, pp. 964–969.

12. I.V. Tsarenko. Adsorptive tetrazole-containing films on steel surface. I. Molecular composition. *Surface,* 1995, No 10, pp. 35–43.

13. R.N. Butler. Recent advances in tetrazole chemistry. *Adv. Heterocyc. Chem.,* 1977, Vol. 21, pp. 323–335.

14. R.D. Holm, and P.L. Donnelly. Spectral studies of metal ion interaction with tetrazole and tetrazolate anion. *J. Inorg. Nucl. Chem.*, 1966, Vol. 28, pp. 1887–1894.

15. D.S. Moore, and S.D. Robinson. Catenated nitrogen ligands. Part II. Transition metal derivatives of triazoles, tetrazoles, pentazoles and hexazine. *Adv. Inorg. Chem.*, 1988, Vol. 32, pp. 171–239.

16. L. Richards, M. La Porte, R. Maguire, et al. Synthesis and characterization of soluble polymeric 5-phenyltetrazolate-bridged aqua cobalt (II) complex. *Inorganica Chemica Acta*, 1978, Vol. 28, pp. 119–122.

17. K. Nakomoto. *IR and CD Spectra of Inorganic and Coordinated Compounds.* NY, Chichester, Brisbane, Toronto, Singapore, John Wiley and Sons, 1986.

18. I.M. Oglezneva, and L.G. Lavrenova. IR spectra in the range of valent vibrations of metal-ligand bond of metal-tetrazole complexes. *J. Inorg. Chem.*, 1985, Vol. 30, No 6, pp. 1473–1478.

19. M.I. Ermakova, I.A. Shikhova, and T.A. Sinitsina, et al. Aminoalkyltetrazoles. II. Complexing of di(5-tetrazolilethylen)oxide, -sulphide and -amine with transition metals ions. *J. Gen. Chem.*, 1979, Vol. 49, No 6, pp. 1387–1391.

20. M.I. Ermakova, I.A. Shikhova, N.K. Ignatenko, and N.I Latosh. Aminoalkyltetrazoles. IV. a,w-Ditetrazolylpolyethylenoxides and -sulphides. Synthesis and complexing with transition metals ions. *J. Gen. Chem.*, 1983, Vol. 53, No. 6, pp. 1364–1368.

21. V.A. Ostrovsky, G.B. Erusalimsky, and M.B. Scherbinin. Study of pentamerous nitrogen-containing heterocycles by quantum chemistry methods. II. Composition and aromaticity of azoles *J. Org. Chem.*, 1995, Vol. 31, No 9, pp. 1422–1431.

22. A.V. Makarevich. Physicochemical and technological principles of thermoplastics based film and fibrous materials creation. Dr. Sci. Thesis, Minsk, 2000.

Chapter 10

1. A. Sasson. *Biotechnologies: Changes and Promises*, 2nd ed. Paris, UNESCO, 1988.

2. Yu.M. Varezhkin, and A.N. Mikhailova. The problem of industrial waste water cleaning in Japan (Review), *J. Mendeleev Sov. Chem. Soc.*, 1991, Vol. 36, No. 1, pp. 79–83.

3. A.V. Putilov, A.A. Kopreev, and N.V. Petrukhin. *Environmental Protection.* Moscow, Khimiya, 1991.

4. S.P. Tsygankov. Waste water utilization of agro-industrial enterprises. *Biotechnology*, 1987, Vol. 3, No.3, pp. 402–407.

5. F. Berne, and J. Cordonnier. *Traitement des Eaux. Epuration des Eaux Residuaries de Raffinage. Conditionnement des Eaux de Refrigeration.* Paris, Editions Technip, 1991.

6. K. Czaczyk, A. Albzecht, A. Mroczkowski, and K. Trojanowska. Mechanical stability of carrageenan and carrageenan/locust bean gum gels used for immobilization of propionic acid bacteria. *J. Biotechnol.*, 1997, Vol. 53, No. 1, pp. 13–20.

7. U.S.S.R. Patent 1291553, C 02 F 3/04. Method of biofilters charging manufacturing. F.V. Shemarov, G.A. Ostretsov, Yu.V. Voronov, et al., 1987.

8. U.S.S.R. Patent 1542917, C 02 F 3/04. Biofilters charging. V.M. Karlovsky, K.V. Kremnev, Yu.V. Voronov, and N.A. Ananjeva, 1990.

9. U.S.S.R. Patent 1560486, C 02 F 3/04. Biofilters charging. Yu.V. Voronov, A.L. Ivchatov, O.B. Netis, and V.P. Solomeev, 1990.

10. A. Khlebikov, and P. Peringer. Biodegradation of p-toluenesulphonic acid by *Comamonas testosteroni* in aerobic countercurrent structured packing biofilm reactor. *Water Sci. Technol.*, 1996, Vol. 34, No. 5–6, pp. 257–266.

11. S.P.P. Ottengraf, M.C.J. Swits, and R.M.M. Diks. Scaling up biofiltration for reliable processes in practice. *Wider Appl. and Diffusion Bioremediation Technol.*, The Amsterdam-95 Workshop, OESD Documents, Paris, 1996, pp. 245–267.

12. F. Kargi, and A. Uygur. Biological treatment of saline wastewater in a rotating biodisk contactor by using halophilic organisms. *Bioprocess Eng.*, 1997, Vol. 17, No. 2, pp. 81–85.

13. R.K. Gorodetskaya, and L.I. Gracheva. Elastic foampolyurethane application as filtering and sorbing material. *Chem. Ind.*, 1991, No.11, pp. 13(653)–18(658).

14. Y.-C. Chung, C. Huang, and C.-P. Tseng. Removal of hydrogen sulphide by immobilized *Thiobacillus sp.* Strain CH 11 in a biofilter. *J. Chem. Tech. Biotechnol.*, 1997, Vol. 69, pp. 58–62.

15. U.S.S.R. Patent 1623982, C 02 F 3/34. Method of water biochemical cleaning from anionic surface-active agents. A.B. Lobova, I.I. Shamolina, S.S. Stavskaya, et al, 1991.

16. K.-H. Engesser, M. Reiser, T. Plaggemeir, and T. Laemmerszahl. Why biofiltration is introduced in indastrial practice? *Wider Appl. and Diffusion Bioremediation Technol.*, The Amsterdam-95 Workshop, OECD Documents, Paris, 1996, pp. 115–121.

17. France Patent Appl. 2639342, C 02 F 3/06. Support de fixation des microorganismes dans lepuration utilisant untel support. G. Valeutis and J. Lesavre, 1990.

18. A.V. Makarevich, L.S. Pinchuk, and I.A. Dunaitsev. New polymer microorganisms carriers in filters for wastewater biological cleaning. *Proc. Belarus AS*, 1997, Vol. 41, No. 1, pp. 114–118.

19. A.V. Makarevich, I.A. Dunaitsev, and L.S. Pinchuk. Aerobic treatment of industrial wastewaters by biofilters with fibrous polymeric biomass carriers. *Bioprocess Eng.*, 2000, Vol. 22, No. 2, pp. 121–126.

20. S.V. Yakovlev, and Yu.V. Voronov. *Biological Filters*. Moscow, Stroyizdat, 1982.

21. Belarus Patent 2753, C 02 F 3/00. Biomass carrier for wastewater biological cleaning filters. A.V. Makarevich, I.A. Dunaitsev, and L.S. Pinchuk, 1999.

22. V.I. Danilov. About magnetic fields influence on biological objects. *Biophysics*, 1990, Vol. 35, No.6, pp. 989–992.

23. M. Kugele, A.B. Yule, and M. Kalaji. An effect of magnetic field exposure on microorganism associated with fuel oil. *Biofouling*, 1999, Vol.14, No.3, pp. 197–211.

24. M. Mehedintu, and H. Berg. New method to evaluate the proliferation response of yeast suspensions on electromagnetic field pyrometers. *Abstr. 8th Eur. Congr. Biotechnol.*, Budapest, 1997, p.193.

25. A.V. Makarevich. Effect of magnetic fields of magnetoplastics on the growth of microorganisms. *Biophysics*, 1999, Vol. 44, No. 1, pp. 65–69.

26. L.A. Musychenko, V.N. Senatorova, L.L. Al'khovskaya, et al. Morphological analysis of microorganism evolution. *Biotechnology*, 1990, No. 3, pp. 3–6.

27. M.V. Volkenshtein. *Physics and Biology*. Moscow, Nauka, 1980.

28. A.G. Alexeev, and Yu.A. Kholodov. Electromagnetic safety. *Bulletin RAEN St.-Petersburg Department*, 1997, Vol. 1, No. 1, pp. 49–54.

29. A.P. Zhukovsky. The biophysical mechanism of the action of magnetic fields and electromagnetic radiation on living organisms. *Abstr. Int. Congr., Weak and Hyperweak Fields and Radiation in Biology and Medicine*, St.-Petersburg, 1997, pp. 47–48.

Chapter 11

1. M.T. Gillies. *Nonwoven Materials.* New Jersey, Noyes Data, 1979.
2. U.S. Patent 4223059, B 29 C 17/02, D 04 H 3/12, D 06 C 3/06. Process and product thereof for stretching a non-woven web of an orientable polymeric fiber. E.C.A. Schwarz, 1980.
3. U.S. Patent 4818464, B 65 H 54/00. Extrusion process using a central air jet. J.C. Lau, 1989.
4. U.S. Patent 4847125, F 16 L 11/02. Tube of oriented, heat shrunk, melt blown fibers. E.C.A. Schwarz, 1989.
5. J. Patockova. Tvorba rouna rozfukovanim taneniny termoplastickeho polymery, *Textil*, 1989, Vol. 44, No. 7, pp. 250–253.
6. U.S.S.R. Patent 612470, B 29 D 7/00. Method of sheet polymer material production. V.P. Shustov, V.A. Belyi, T.V. Stavrova, and O.R. Yurkevich, 1978.
7. C. Dickenson, ed. *Filters and Filtration. Handbook.* 3rd ed. Oxford, Elsevier, 1992.
8. U.S.S.R. Patent 533110, H 01 M 6/00. 10/00, 2/16. Method of chemical current source production. L.S. Pinchuk, A.S. Neverov, and V.A. Goldade, 1974.
9. U.S. Patent 4526733, B 29 F 3/04. Melt-blown die and method. J.C. Lau, 1985.
10. V.N. Kestelman, L.S. Pinchuk, and V.A. Goldade. *Electrets in Engineering. Fundamentals and Applications.* Boston, Dordrecht, London, Kluwer Academic, 2000.
11. U.S. Patent 5186879, B 29 C 47/88. Spinning process for producing high strength, high modulus, low shrinkage yarns. F.H. Simons, and R.L. Griffis, 1993.
12. U.S. Patent 4380570, D 04 H 1/04. Apparatus and process for melt-blowing a fiber-forming thermoplastic polymer and product produced thereby. E.C.A. Schwarz, 1983.
13. U.S. Patent 5476616, B 29 C 47/30. Apparatus and process for uniformly melt-blowing a fiber-forming thermoplastic polymer in a spinneret assembly of multiple rows of spinning orifices. E.C.A. Schwarz, 1995.

Chapter 12

1. V.I. Vernadsky. *Philosophic Thoughts of a Naturalist.* Moscow, Nauka, 1988.
2. R. Kurane. *Bioprevention of air pollution*, OECD Documents: Wider application and diffusion of bioremediation technologies. The Amsterdam'95, Workshop, Paris, OECD, 1996, pp. 269–279.
3. I.I. Mazur, and O.I. Moldovan. *Introduction to Engineering Ecology.* Moscow, Nauka, 1989.
4. C. Dickenson, ed. *Filters and Filtration. Handbook*, 3rd ed. Oxford, Elsevier, 1992.
5. T.V. Stavrova. Elaboration and investigation of fibrous-porous materials produced by polymer melt spraying. Ph.D. Thesis, Minsk, 1983.

6. 3M petroleum sorbents. St.Paul, 3M Center, 1996.

7. Advanced sorbents technology. Los Angeles, SPC Sorbent Products Co., Inc., 1995.

8. U.S.S.R. Certificate of industrial specimen 26863. Universal rotary mincer. V.A. Rozhnov, S.Ya. Liberman, G.A. Goncharov, et al., 1988.

9. U.S.S.R. Patent 1567659, D01G 1/04. Arrangement for non-oriented fiber cutting. Yu.I. Pushkarev, P.P. Chueshkov, and A.A. Rodin, 1990.

10. U.S.S.R. Patent 1484482, B23D 25/00. Arrangement for material cutting and method for its knife sharpening. Yu.I. Pushkarev, A.E. Yunitsky, Yu.M. Pleskachevsky, et al., 1989.

11. U.S.S.R. Patent 943344, D01G 0/4. Arrangement for fibrous material cutting. N.M. Klimashevich, Yu.I. Pushkarev, M.A. Demus'kov, et al., 1982.

12. U.S.S.R. Patent 1317041, D01G 0/4: Arrangement for fibrous material cutting. Yu.I. Pushkarev, A.E. Yunitsky, D.A. Emanov, 1987.

13. U.S.S.R. Patent 996536, D01G 0/4. Arrangement for fibrous material cutting. S.V. Scherbakov, N.M. Klimashevich, Yu.I. Pushkarev, et al., 1983.

14. U.S.S.R. Patent 1187879, B02C 18/06, 18/44. Arrangement for milling. S.Ya. Liberman, A.E. Yunitsky, A.A. L'vov, and Yu.I. Pushkarev, 1985.

15. D. Loveland. Advancing the development of biodegradable polymers, OECD Documents: Bioremediation. The Tokyo'94 Workshop, Paris, OECD, 1995, pp. 609–614.

16. E.G. Lyubeshkina, ed. *Recycling of Polymer Materials*. Moscow, Khimia, 1985.

17. A.V. Makarevich. Effect of magnetic field of magnetoplastics on the growth of microorganisms. *Biophysics*, 1999, Vol. 44, No. 1, pp. 65–69.

18. A.V. Makarevich, I.Yu. Ukhartseva, V.A. Goldade, and L.S. Pinchuk. Self-decomposing polymer packaging materials. *Plast. Masses*, 1996, No.1, pp. 34–36.

Subject Index

Acid benzoic 13
Additives active 8
Adhesion 12
Adsorbents
– inorganic 135
Adsorption
– index 137
– low-temperature 58
– of oil products 138
Aerator pneumatic 181
Aerosil 135
Aerosol
– highly dispersed 95
– polymer-solvent 12
Agents
– complexing 135, 139
Agglomerates 117
Aggregates 16, 66–68, 119, 124, 126, 131
– destruction of 17
– stability 127
Air
– fine purification 106
– ventilation 106
Aluminosilicate enamel 132
Antiadhesive 18
Antibacterial effect 144
Antibiotics 154, 155
Articles
– biodegradable 163
– fluffy 163
– miscellaneous textile 164
– with characteristic trimming 164
Attraction magnetic 16

Bandages
– nonwoven spun 172
Benzene 137, 184
Biax -Fiberfilm Corp. 43
Biocompatibility 148, 156

Biofilm 148, 152–154
Biofilter 147, 149 154, 181
Biomass 147, 152, 158
Bioreactor 148, 158
Bond cohesive 5
Breaking stress under tension 112
Brownian motion 88
Burial 184, 187

Capability
– petroleum-retaining 137
Capacity
– adsorptive 139, 140, 153
– specific sorptive 150
Capture
– efficiency of 114
– magnetic 119
– mechanical 109
– of contaminants 127, 180
– of particle 107
– section 97
Carbonyl iron 117, 132
Carpets 165
Carriers 150–155, 158
– fibrous polymer 148
– magnetic 154, 158
Characteristics
– adsorptive 136
– service 161
Charge
– carriers 14
– electret 77, 80, 95
– polarizing 14, 80
– spontaneous electric 75
Charging
– two-side 103
Chemical current sources 167
Chemical oxygen demand 150
Clathrate hydrates 130
Claydite 149- 155

Cleaning
- biological 147–149, 152, 183
- degree of 83
- efficiency 155, 183
Clothes disposable 163
Coagulation 113
- acoustic 93
- magnetic 93
Coal
- activated 135
- active 135
Coalescence 20
Coatings
- antistatic frictional 174
- electret 95, 97
- fibrous 169
- heat- and soundproofing 169
Components
- complexing functional 143
- functional 138
Compounds
- coordination 139–142
- nitrogen-containing heterocyclic 135, 139
Compression 168
Condensation
- capillary 99, 100
Conductivity
- electron 122
- equilibrium 98
- ionic 122
Containers and bags 174
Contaminants 83, 88
- artificial 85
- bacterial 144
- capture of 127
- ferromagnetic 115
- organic 19
Conveyor belts 168
Coulomb's low 90
Crown esters 135

Davies's equation 114
Depolarization
- thermal 98
- thermally stimulated 76–81
Depolymerization 185, 187
Derjaguin–Landau theory 101
Destruction
- bio- 181
- degree of 31
- of polymer 10
- thermal oxidative 11, 65, 73, 98

Diffusion 121, 143
Dioctyl phthalate 121
Discharge
- corona 89, 95, 104, 122
- high-voltage 7
Disinfection 143
Dissolved
- ions 74
- metals 72
- SrO 73
Dithizone 13
Double electrical layer 101, 123
Dupret–Young equation 102

Effect
- antibacterial 144
- electrocapillary 101
- magnetic electrical 124
- magnetobiological 158
- Pool–Frenkel 99
- screening 85, 119, 121
Electrets
- corona 97
- thermo- 102
Electrification 77
- tribo- 89
Electrocapillary effect 101
Electrodes
- corona 15
Electron paramagnetic resonance 158
Elements
- aerating fibrous 19
Endoprostheses
- vessel 171
Energy
- activation 97
- magnetic interaction 113
- precipitation 113
- surface 102
Engineering ecology 179

Fabrics
- for cleaning and polishing 163
- protective 165
Facings
- relief fibrous 169
Ferrite
- agglomerated magnetized particles 117
- barium 15, 118, 133
- concentration 15, 156
- dissolution of 71
- single-domain particles 113, 117

– strontium 15, 61, 68, 118
– X-ray electron spectra 70
Ferrocyanides 135
Fibers
– carbon 135
– density 6, 53
– deposition 8
– diameter 6, 53, 60, 105
– distribution 53, 55, 57
– flow 37
– from polymer solution 45
– mechanical entanglement of 7
– mixture of 6
– modification of 36, 49
– stretching and breakage 22
– vacancies between 59
Fibrous fabrics
– biaxially extending of 48
Field
– electrostatic 14
– magnetic 15, 111, 121
– high-gradient magnetic 127
– electromagnetic 16
– texturing 16, 121
– packing density 17
– high-voltage 103, 104
– inhomogeneous magnetic 121
– of corona discharge 122
– gradient of intensity 122
– physical 10, 13
– electrical 14, 44, 104
Filler
– magnetosolid 17
Filtering factor 83
Filters
– bio- 147, 149–151, 152
– cake-like (filled) 89
– electret 89
– paper 133
– porous 88
Filtration
– depth 59
– efficiency 89, 105, 108, 115, 138, 149, 150
– fineness 84
– mineral oil 116
– ratio 84
– water 116
Floors
– asphalt 168
– decorative 169
– for gym-halls 168
Foodstuff packages 175

Forces
– adhesive 91
– adsorptive 86
– capillary 100103
– centrifugal 23, 24
– coulombic 89, 131
– inductive 89
– inertial 24
– Lorentz 93, 127, 131
– magnetic 92, 123
– mechanical 126
– of magnetic attraction 115
– surface tension 99, 102, 126
– van der Waals attraction 131
Friction
– internal 17

Gas flow
– pulse 35

Heat exchange 30

Impaction
– internal 85, 121
Impurities
– iron-containing 91 iron 111
Incineration 185, 186
Induction
– magnetic 17, 118–120, 155
Insulation
– electrical 166
– soundproofing 166
– heat 166
Interaction
– electron 141
– electrostatic 103
– magnetic 113
– molecular (van der Waals) 101, 132
– spatial 141
Interception
– direct 86
– inertial 118, 121, 132
– metal ion 143
Ionites 135
Ions
– heavy metal 12

Larmor precession 130
Leather substitutes 164
Liquid
– diffusing and dissolving 103
– silicone-organic 13
– surface tension 129

Magnetic electrical effect 124
Magnetization 16, 113
– axial 157
– spontaneous 128
Magnetobiological effect 158
Magnetoplastics 155–158
Materials
– binding 167
– biologically active 147
– camouflage 176
– cleaning and polishing 175
– density 62, 87, 88, 105, 112
– electret 180
– for curing burns 172
– for marine containers 174
– for tight containers 174
– hindering soil erosion 176
– insecticidal 177
– lining 170
– nonwoven 6, 46, 173
– packing 173
– packing density 104, 133
– protein 173
– specific area 53, 57, 59
– surgical 170
– textile 163
– to protect wood 177
– vacancy size 57
– wiping 175
– woven 6
Mats
– for collecting petroleum 176
– load-carrying 169
– supporting 169
– waterproof 169
Melioration 2
Melt index 66
Melt-blown
– disposables 161
– strengthening of articles 32
– tubular articles 44
Melt-spinning 50
Meshes
– camouflage 177
– of paper-making machines 167
– radio-absorbing 177
Metabolism 183
Methylene blue 137
Microbial cell 156
Microorganisms 145–149, 152, 155–157
– carriers of 147, 150
– complex cultures 149
– immobilization 149, 158

– suppression 145
Modifier
– magnetosolid 16
Molecular mass 11
Mud capacity 85, 180

Napkins
– for burnt wounds 173
– for furniture polishing 175
– hygienic 162
– sanitary 163
– with a rubber foam coating 175
Needle-punching 7
N-methylolacrylamide 165

Oil
– lubricating 111
– petroleum 138
– products 152
– vaseline 13
Optical-magnetic detector 117, 122

Paraffin 120
Particles
– agglomerated magnetized ferrite 117
– aggregates 124, 126 ferrite 128
– diameter 108
– distribution of 68
– ferromagnetic contaminants 115, 119, 131
– magnetosolid 17
– radioactive 1
– single-domain ferrite 113, 117, 120
– solid 9, 36, 114
– submicron size 99, 180
Penetration
– capillary 88
Petri dishes 144
Petryanov's filter 103, 107
Pipes
– clogging with silt 182
– drainage 18
– filling 41
– melt-blown 18
– perforated 22
Plasticizer 8
Polarization
– current density 96
– dipolar 105
– electrical 13, 38
– low-temperature 98
– Maxwell–Wagner 98, 99, 105

– mechanism 79
– volume-charge 105
Polyacrylate 164, 168, 177
Polyacrylonitrile 135
Polyamide 61, 66, 112, 117, 133, 135,
    159, 164, 166, 168, 170, 175, 177
Polyarylate 190
Polyester 168, 170, 175, 177, 180, 190
Polyethylene 112, 117, 159, 174, 189,
    190
– films 67, 69, 174
– high-density 105
– low-density 61, 66, 80, 95, 128, 186
Polymers
– blends 105
– modification 154
– oxidation 66–68, 72
Polyolefin 170, 174, 175
Polypropylene 89, 107, 168, 170, 177,
    180
Polyurethane 164, 165, 168, 170
Polyvinylacetate 135, 172, 176
Polyvinylchloride 105, 164, 168, 172,
    177
Pool–Frenkel effect 99
Porcelain 132
Pores
– meso- 136
– micro- 136
Porosity 14, 112
– general 53–55, 59
– gradient 18, 87
– of fibrous matrix 18
Potential
– electrical 14
Powders
– components 37
– ferromagnetic 7, 15
– nickel 121
Precipitation
– diffusive 86
– energy 113
– gravitational 85, 118, 121, 132
– interceptive 113
– magnetic 113
Preparations
– microbicidal 143
Pressure
– differential 84, 89, 132
– wedge (disjoining) 101
Products elastics 164
Protection
– against biological injury 177

Purification
– gases 180
– water 130
Pyrolysis 185, 187

Quartz sand 132

Radiation
– infrared 2
– ionizing 105
– ultraviolet 174
Receptivity
– magnetic 125
Recycling 185
Reicofil 43
Reifenhäuser 50
Relative elongation at rupture 112
Relaxation
– heterocharge 96
– homocharge 96
– time 96, 99
Resistance
– aerodynamic 105 108
– hydrodynamic 89, 115
– specific 90
– Stokes 131
– to dynamic tension 168
Respirators 106
Reverse corona 91
Rubber 168
– -silicone membranes 172

Saturation
– magnetic 16
Screening effect 85, 118, 121
Sedimentation
– zone of 123
Serviettes surgical 170
Sewage 147
Sheets absorptive 176
Skipping coefficient 84, 105, 108
Sorption
– specific 136, 149
– volume of 136
Spectra
– X-ray electron 70, 142
Splints
– water-curable 172
Spray heads 31
– multichannel 25, 26, 32
– rotary 23
– single-channel 21, 22
– slot 27

– temperature of 29 original 39
– with numerous spinneret holes
    27, 33
Spreading 102
Spunbonding 7
Stability
– dimensional 2
– of aggregates 127
– of electret 99
– structural 88
State
– charge 77–81
– electret 15
– viscous-flow 8, 121
Stokes
– coefficient 123
– low 93
– resistance 131
Stokes–Kanningham's low 90
Structure
– crystalline 141
– fibrous 56, 95
– fibrous-porous 53, 54
Substances
– adsorptive 135
Swirler 22

Temperature
– control 28
– gradient 18
Tetrachloride carbon 13
Tetrazoles 139, 143, 144
– ligand 142
Thermoelastoplast 170
Thermoplastics 175, 177
Threebutyl phosphate 13
Towels 161
– surgical 171
Transmission (skipping) factor 84

Trapping
– centers 97, 99
Treatment
– magnetic 121, 130
– magnetic of aqueous solutions 131
– magnetic of water 129
– supersonic 7
– thermal 7
– ultrasonic 121

Vibration 22
Viscosity
– of melt 11

Wallpapers
– sound-insulating 169
Wastes 150
– chemical 151
– domestic 186
– plastic 186
– production 185
– technological 185
Water
– crystalline 142
– magnetic treatment 129
– waste- 149, 151–155, 179, 181
Wear
– debris 111
– metal debris 134
Wetting 99, 102
– angle 129
– edge angle 101
Work
– adhesive 102
– cohesive 102

Young's low 102

Zeolites 135

Springer Nature Customer Service Center GmbH
Europaplatz 3, 69115 Heidelberg, Germany

Printed by Elias Plösser GmbH
in Hamburg, Germany